Multimedia and Interactive Digital TV: Managing the Opportunities Created by Digital Convergence

Margherita Pagani
I-Lab Research Center on Digital Economy
Bocconi University, Italy

IRM Press
Publisher of innovative scholarly and professional
information technology titles in the cyberage

Hershey • London • Melbourne • Singapore • Beijing

Acquisitions Editor: Mehdi Khosrow-Pour
Senior Managing Editor: Jan Travers
Managing Editor: Amanda Appicello
Copy Editor: Maria Boyer
Typesetter: Amanda Appicello
Cover Design: Weston Pritts
Printed at: Integrated Book Technology

Published in the United States of America by
    IRM Press (an imprint of Idea Group Inc.)
    701 E. Chocolate Avenue, Suite 200
    Hershey PA 17033-1240
    Tel: 717-533-8845
    Fax: 717-533-8661
    E-mail: cust@idea-group.com
    Web site: http://www.irm-press.com

and in the United Kingdom by
    IRM Press (an imprint of Idea Group Inc.)
    3 Henrietta Street
    Covent Garden
    London WC2E 8LU
    Tel: 44 20 7240 0856
    Fax: 44 20 7379 3313
    Web site: http://www.eurospan.co.uk

Copyright © 2003 by IRM Press. All rights reserved. No part of this book may be reproduced in any form or by any means, electronic or mechanical, including photocopying, without written permission from the publisher.

Library of Congress Cataloging-in-Publication Data

Pagani, Margherita, 1971-
    Multimedia and interactive digital TV : managing the opportunities created by digital convergence / Margherita Pagani, PhD.
        p. cm.
Includes bibliographical references and index.
    ISBN 1-931777-38-1
 1. Interactive television. I. Title.
    TK6679.3 .P34 2003
    384.55--dc21
                                                    2002153244

eISBN: 1-931777-20-9

British Cataloguing in Publication Data
A Cataloguing in Publication record for this book is available from the British Library.

New Releases from IRM Press

- **Multimedia and Interactive Digital TV: Managing the Opportunities Created by Digital Convergence**/Margherita Pagani
  ISBN: 1-931777-38-1; eISBN: 1-931777-54-3 / US$59.95 / © 2003
- **Virtual Education: Cases in Learning & Teaching Technologies**/ Fawzi Albalooshi (Ed.), ISBN: 1-931777-39-X; eISBN: 1-931777-55-1 / US$59.95 / © 2003
- **Managing IT in Government, Business & Communities**/Gerry Gingrich (Ed.)
  ISBN: 1-931777-40-3; eISBN: 1-931777-56-X / US$59.95 / © 2003
- **Information Management: Support Systems & Multimedia Technology**/ George Ditsa (Ed.), ISBN: 1-931777-41-1; eISBN: 1-931777-57-8 / US$59.95 / © 2003
- **Managing Globally with Information Technology**/Sherif Kamel (Ed.)
  ISBN: 42-X; eISBN: 1-931777-58-6 / US$59.95 / © 2003
- **Current Security Management & Ethical Issues of Information Technology**/Rasool Azari (Ed.), ISBN: 1-931777-43-8; eISBN: 1-931777-59-4 / US$59.95 / © 2003
- **UML and the Unified Process**/Liliana Favre (Ed.)
  ISBN: 1-931777-44-6; eISBN: 1-931777-60-8 / US$59.95 / © 2003
- **Business Strategies for Information Technology Management**/Kalle Kangas (Ed.)
  ISBN: 1-931777-45-4; eISBN: 1-931777-61-6 / US$59.95 / © 2003
- **Managing E-Commerce and Mobile Computing Technologies**/Julie Mariga (Ed.)
  ISBN: 1-931777-46-2; eISBN: 1-931777-62-4 / US$59.95 / © 2003
- **Effective Databases for Text & Document Management**/Shirley A. Becker (Ed.)
  ISBN: 1-931777-47-0; eISBN: 1-931777-63-2 / US$59.95 / © 2003
- **Technologies & Methodologies for Evaluating Information Technology in Business**/ Charles K. Davis (Ed.), ISBN: 1-931777-48-9; eISBN: 1-931777-64-0 / US$59.95 / © 2003
- **ERP & Data Warehousing in Organizations: Issues and Challenges**/Gerald Grant (Ed.), ISBN: 1-931777-49-7; eISBN: 1-931777-65-9 / US$59.95 / © 2003
- **Practicing Software Engineering in the 21st Century**/Joan Peckham (Ed.)
  ISBN: 1-931777-50-0; eISBN: 1-931777-66-7 / US$59.95 / © 2003
- **Knowledge Management: Current Issues and Challenges**/Elayne Coakes (Ed.)
  ISBN: 1-931777-51-9; eISBN: 1-931777-67-5 / US$59.95 / © 2003
- **Computing Information Technology: The Human Side**/Steven Gordon (Ed.)
  ISBN: 1-931777-52-7; eISBN: 1-931777-68-3 / US$59.95 / © 2003
- **Current Issues in IT Education**/Tanya McGill (Ed.)
  ISBN: 1-931777-53-5; eISBN: 1-931777-69-1 / US$59.95 / © 2003

Excellent additions to your institution's library!
Recommend these titles to your Librarian!

To receive a copy of the IRM Press catalog, please contact
(toll free) 1/800-345-4332, fax 1/717-533-8661,
or visit the IRM Press Online Bookstore at: [http://www.irm-press.com]!

Note: All IRM Press books are also available as ebooks on netlibrary.com as well as other ebook sources. Contact Ms. Carrie Stull Skovrinskie at [cstull@idea-group.com] to receive a complete list of sources where you can obtain ebook information or IRM Press titles.

*To my parents and to the ones who taught me to see how
large is the world and how wide are the horizons*

Multimedia and Interactive Digital TV:
Managing the Opportunities Created by Digital Convergence

Table of Contents

Preface x

PART I: DIGITAL CONVERGENCE

Chapter I
The Digital Metamarket 1
    The Emerging Multimedia Metamarket 1
    The Key Drivers of Convergence 4
    The New Emerging Markets 8
    The Organization of the New Value Chain 12
        The New Value Chain of the Television Industry 15
            Content Providers 15
            TV Channels 15
            Multi-channel Network 16
            Satellite Platforms 16
            Cable Operators 16
            Terrestrial Broadcasters 16
            Satellite Operators 17
        Alliances at Different Levels of the Value Chain 17
            Horizontal Integration Strategies 19
            Vertical Integration Strategies 22
    Evolution of the Offer 25

Chapter II
New Digital Media and Devices: Measuring the Potential for IT Convergence at Macro Level 31
   Introduction 31
   Convergence Discussed 33
      A Classification of New Digital Media and Devices 33
         Transport Media 34
         End Devices 35
         The Matrix of Transport Media and End Devices 35
      Barriers to Convergence 37
   A Definition Based on Platform Penetration and CRM Potential 41
      The Concept of "Critical Digital Mass Index" 42
      The Convergence Factor 43
      The Relevance of Interactivity 46
      The Convergence Index 47
   Conclusions 48

PART II: MULTIMEDIA AND INTERACTIVE DIGITAL TELEVISION

Chapter III
Digital Television 53
   Introduction 53
   What is Digital Television? 54
      Digital Coding 54
      Digital Compression 55
   Advantages Offered by Digital Television 57
   Transmission Systems 59
      Digital Satellite Transmission Systems 60
      Digital Cable Transmission 61
      Digital Terrestrial Television (DTT) 62
      ADSL (Asymmetric Digital Subscriber Line) Transmission 63
      MMDS (Multichannel Multipoint Distribution System) 65
      LMDS (Local Multichannel Distribution System) 67
   Reception Systems 67
      General Functions 68
         Conditional Access 69
         Return Path 70
         Electronic Programming Guide 70
   The Problem of Standards 71

Chapter IV
The Economic Implications of Digital Technologies 75
    Advantages and Limits in Digital Signal Transmission Methods 75
        Digital Terrestrial Television 75
        Television Broadcast Via Satellite 77
        Television Broadcast Via Cable 78
        Television Broadcast Via ADSL (Asymmetric Digital Subscriber
            Line) 79
        Television Broadcast Via Microwave MMDS (Wireless Cable). 80
        Comparative Analysis of Different Transmission Technologies 82
    The Economic Effects of Digital Transmission 85
        The State 85
        The Viewer 86
        The Operators Involved 89
    Costs and Penetration Time Compared 90

Chapter V
Interactive Digital Television 96
    Introduction 96
    A Definition of Interactivity 97
        Local Interactivity 100
        One-Way Interactivity 100
        Two-Way Interactivity 101
    'Interactivity': Prototype, Criteria, or Continuum? 102
    Capacity: Downstream and Up-Stream Bit-Rates 110
    Interactive Television 111
    New Offer Types 112
    Electronic Program Guide and Pay-Per-View 117
        Electronic Program Guide 117
        Pay-Per-View 118
    Interactive TV Services 118
        Interactive Games 119
        Interactive Advertising 120
        TV Shopping 121
        TV Banking 122
        Model of Interactive Programming 123
    Interactive Digital Television (iDTV) Value Chain 125
    Becoming a Customer or Content Gateway in the New Economy 128

PART III: MANAGING THE OPPORTUNITIES CREATED BY DIGITAL CONVERGENCE

Chapter VI
Branding Strategies for Digital Television Channels 135
    New Sources of Competitive Advantage Become Established 135
    The Customer-Based Brand Equity Concept 136
    Competition Pressure Increase 137
        Digital Television Networks Types 138
        Competitive Mechanisms 140
    Launching of a Themed Channel on the Italian Television Market:
        The Disney Channel Case 142
    Viewers' Cognitive System Analysis Parameters 144
    Channel-Watching Motivation Trigger 147
    Television Brand 148
    Brand Communication Tools Available to Digital Television
        Channels 150
    Interactive Portal Tools Available to Increase Viewers' Loyalty 152
    Conclusions 153

Chapter VII
The Critical Role of Content Media Management 156
    Introduction 156
    Definition of Content Management 158
    Managerial Implication for the Digital Broadcaster 161
    Business Evolution Through Digital Enablement 165
        Strategy 167
        Enablement 167
        Integration 168
        Transformation 168
    The Migration Toward Digital TV 168
    The Mediaset Experience in Italy 170
        Mediaset Innovation Projects 174
            Digital TV Innovation Project in the Production Area 174
            Digital TV Innovation Project in the Contribution Area ... 176
            Digital TV Innovation Project in the Archiving Area 177
            Digital TV Innovation Project in the On-Air Area 177
    Conclusions 178

Chapter VIII
Digital Rights Management 180
    Introduction 180
    Intellectual Property: A Definition 181
    Key Intellectual Property Rights Issues 182
        Ownership in the Digital World 183
        Distribution 183
        Protection of Intellectual Property 184
        Globalisation 184
        Standards 185
    Digital Rights Management: Functional Architecture 185
    Digital Rights Management: Value Chain Activities 189
    Digital Rights Management Benefits 193
    Conclusions 194

Chapter IX
Conclusions 196

Appendix 201
    Acronyms and Abbreviations 201
    Glossary 203

References 218

About the Author 233

Index 234

Preface

In the past few years, managers active in the multimedia value chain have been confronted with new phenomena having a remarkable impact on corporate management. These phenomena are caused by the development of sectors such as: technology dynamics, digital economy, demand development, liberalisation, and deregulation policies, as well as a general increase in competitive pressure.

Therefore, digital convergence does not refer to a single event, but rather to the occurrence of an evolutionary process through which different sectors and technologies—which were originally more or less independent from each other—merge.

The whole television industry is involved in this ongoing convergence process, and new challenges and opportunities are opening up to organisations.

The impact of digital technology is the first determinant in evaluating the technological asset as the origin of the structural and competitive transformation of the television market. The technological change brought about by the advent of the new digital technologies will be surveyed here, showing the significant economic components from a relativistic approach.

Digital television is based on the transmission of a numerical or digitised signal that is transformed through algorithms into a signal that removes all redundancies of space and time. The signal is broadcast through compelling delivery systems (cable, satellite, terrestrial, optical fibers) The user's reception of the digital signal is made possible through a digital adapter (*set top box* or *decoder*) which is connected to the normal television or integrated with the digital television in the latest versions.

At present, in Europe more than 23 million households are equipped with digital television, but by 2005, more than 64 million subscribers in Europe and 46 million subscribers in the U.S. are expected to have access to digital television. The expected growth is very considerable.

The advantages offered by digital television to the viewer include:

- an increase in the number of channels that can be transmitted (because of a reduction in the use of the electromagnetic spectrum due to the compression of the digital signal);
- better transmission of image and sound quality;
- the possibility to use larger format; and
- television screens (from 16:9 to large size flat screens).

The second set of factors relates to the series of additional interactive services that can be added to television, and that allow users' interaction with the offered content. For this reason, it is worthwhile to distinguish between the digital and the interactive concepts, as the latter is made possible by the existence of new digital technologies.

Changing competitive systems (characterized by an increasing number of television channels, due to technological innovation in the transmission of digital signal) and the convergence process among sectors, which in the past were distinct and separated, are causing a considerable impact on competitive dynamics and a constant review of media companies' competition patterns.

To be active within hypercompetitive markets, an organisation must develop a set of skills which are characteristic of a proactive organisation.

Technological innovation causes media companies to face a new paradigm: either self-transformation or the risk to find itself in an "electronic quagmire" by new technologies.

Convergence is not solely a technological matter; it's a brand new life and working style, as it foresees new services and opportunities to implement industry productivity and competitiveness on the market. Media, information technology, and telecommunications are then to take advantage of new products and platforms to become part of a unique global network. Digital communications are the first and essential step to the convergence of information and telecommunications technology where traditional media, once clearly distinct and independent from each other, meet in the new land of interactivity and multimedia.

In this book, the impacts generated by the shift from digital technology and interactivity development within the TV industry are investigated, with a focus on some corporate management implications. The focus of experience covered in this book is European. A few basic assumptions are made.

The first assumption relates to the business backbone: boundaries of each single sector (telecommunications, IT, TV, electronics) which can be detected within the multimedia metamarket become less and less definite and increasingly overlapping. Therefore, the ongoing convergence process must be investigated.

The concept of digital convergence is broken down into the three following levels: convergence of devices, convergence of networks, and convergence of content. Although there is evidence in digital environments of limited alignment in some of these areas, there are considerable physical, technical, and consumer barriers in each case.

Rather than convergence, the transition from analogue to digital is often being accompanied by a process of fragmentation. For this reason, it is suitable to trace the limits of convergence, understood as a phenomenon that involves the bringing together, merging, or hybridisation of different types of digital device, network, or content.

A second assumption makes reference to the peculiar features of digital TV and different transmission modes of the digital signal. Each broadcasting modality (cable, satellite, terrestrial, ADSL, MMDS) shows specific advantages and constraints which any medium-size organisation must take into account in order to set out a correct strategy.

Based upon the above, openings to digital convergence development and its managerial applications for TV companies into the new competitive arena must be discussed.

Among the corporate intangible assets, undoubtedly brand and trust play a critical role, and for this reason the critical success factors for a TV organisation in the new digital era appear to be connected with the development of careful branding policies in order to acquire and make their audience loyal.

Another critical role is also played by content asset management, which is the ability to manage a digitalised and stored audio-video content and distribute it on different platforms.

In summary, this book contributes to analysing a complex and changing business in the digital era by listing critical points as well as opportunities for the major organisations.

The rules of the game, the reference competitive scenario, the viewers' habits, and offering modalities are all undergoing big changes, and at present, new business models are available to networks coupled with endless opportunities for development and growth within this increasingly fascinating and complex business.

Book's Structure

To achieve a better understanding of this book, its first part is devoted to investigating the new boundaries of the TV networks for the purpose of identifying the system pivotal points and defining an easy conceptual model, based upon the definition of digital metamarket.

Components of the multimedia value chain and convergence drivers are analysed and major players and their strategies in the different stages of the multimedia value chain are studied (Chapter 1). In approaching the study of the so-called multimedia metamarket backbone, which is where a TV network operates, it is crucial to preliminarily clarify its components and functions.

In Chapter 2 the concept of digital convergence is used to refer to three possible axes of alignment: convergence of devices, convergence of networks, and convergence of content. Although there is evidence in digital environments of limited alignment in some of these areas, there are considerable physical, technical, and consumer barriers in each case. In fact, rather than convergence, the transition from analogue to digital is often being accompanied by a process of fragmentation.

A better way of looking at convergence may lie in the degree to which two-way digital networks facilitate cross-platform management of customer relationships—regardless of the type of networks those customers inhabit or the kind they consume. Chapter 2 argues for a definition of convergence based on penetration of digital platforms and the potential for cross-platform Customer Relationship Management (CRM) strategies, before developing a convergence index according to which different territories can be compared.

Once the conceptual system is defined and a terminology in line with the study of the subject matter is developed, digital and interactive TV are discussed in the second part of this book.

First, the technological area is discussed in detail through the features of digital TV and the economic implications generated by the different signal broadcasting technological modalities (Chapters 3 and 4). The development of interactive television is a further evolution phase of the digital television through new functionalities and interactive applications which make richer offerings through the development of an increasingly more customised interaction with final viewers (Chapter 5).

In the third part of the book, the impact on corporate management from new opportunities based upon the development of digital convergence is discussed in greater detail. More specifically, the following three managerial areas influenced by these new opportunities were detected:

- growing importance of brand equity as a loyalty resource available to digital TV channels to aggregate and make a more concrete stand vis-à-vis viewers' loyalty (Chapter 6);
- the critical role of the content asset management or digitalisation of audio and video contents, their management in digital libraries, and their distribution through different digital platforms (Chapter 7);
- the new role of Digital Rights Management processes in the typical digital media management value chain and the security issue in a digital TV world (Chapter 8).

In this phase of the study, technological scenarios within possible models have tentatively been identified, particularly those which have a greater likelihood of being successful. Interactive television development and its services to viewers, as well as its key management implications, have also been discussed.

The research work is based upon direct contributions from interviews with people in this business. A detailed analysis on the global technological and competitive scenarios was carefully made through desk research and direct contacts with the major corporate entities in this business.

The study on interactive television was first discussed at the symposium on the emerging electronic markets held in Muenster in September 1999; a more advanced phase of this study was also discussed during the Business Information Technology Conference (BIT) in Mexico City in May 2000, as well as at the SISEI Conference held at Bocconi University in December the same year. Chapter 7, "Critical Role of Content Management," was discussed during the IRMA 2001, International Resource Management Conference in Toronto in May 2001, and Chapter 2 was discussed during the IRMA 2002, International Resource Management Conference in Seattle in May 2002.

The above academic conferences have all been a very good chance to check and compare data and assumptions, thus providing this study with a remarkably richer perspective.

Precious contributions were also made by expert researchers from some of the major European and U.S. universities.

Acknowledgments

This book is the outcome of a long research project running for several years to look into the future of multimedia and interactive digital TV. Much of the book is based on findings of a four-year multi-client project on digital television and multimedia convergence realised at the New Media&TV-Lab, a research laboratory inside the I-LAB Research Center on Digital Economy of Bocconi University in Milan, Italy.

The research work was supported and made possible, particularly in its initial stage, by Bocconi University as well as by sponsorships and contributions from the major Italian media companies which deserve my most heartfelt thanks.

The creation of the New Media & TV-Lab inside the I-LAB Research Center on Digital Economy of Bocconi University in the year 2000 had the goal of studying the TV industry and the convergence phenomena between traditional and new media. The laboratory, as well as its research activities, were the fundamental bases of this book.

My heartfelt thanks also to those organisations which encouraged and helped me with the New Media&TV-lab during these years, namely 24 Ore Television, Disney Channel, Home Shopping Europe, HDPnet, Motorola, Mediaset, SHS Multimedia, MTV Network, Sitcom, Sony, Stream, Telepiù, Nokia, and Philips.

In the course of this journey, I have interviewed scores of people, listened to experts, and gauged public opinion on interactive digital television. I have discussed my findings and the content of this book with participants and delegates in seminars, lectures, workshops, and international academic con-

ferences in the U.S. and Europe. To all those who have contributed with views, suggestions, criticism, and encouragement, I wish to extend my thanks.

My most sincere thanks also to all those who have made this work possible, first among them all, Professor Enrico Valdani (Director of I-LAB), who deserves all my admiration and deepest gratitude not only for his role of guide and for the endless pieces of advice, but especially for having taught me how important commitment and determination are to make one's own ideals come true.

My most sincere thanks also to all my friends and colleagues (in particular, Vanessa Manzione for the research assistance provided through these years) and to all those who have encouraged me with their appraisal and affection.

I'm very grateful to several people for reading significant pieces of the manuscript or an entire draft of this book, and for offering valuable advice on how it could be improved. They include also the three anonymous reviewers who gave me a number of ideas, as well as very useful feedback on the entire draft of this book. Special thanks to Paolo Noseda for his help with the translation of the manuscript and Amanda Appicello at IGP for her precious assistance during all steps of the publishing.

Finally, I would also like to thank my family for their unwavering support both throughout the writing of this book and over the years.

Margherita Pagani
Milan, August 2002

PART I

DIGITAL CONVERGENCE

Chapter I

The Digital Metamarket

THE EMERGING MULTIMEDIA METAMARKET

The world's emerging multimedia market results from the process of convergence of three industries which were created at an interval of 50 years respectively—the telephone industry (1890), the television industry (1930), and the computer industry (1980).

Convergence describes a process change in industry structures that combines markets through technological and economic dimensions to meet merging consumer needs. It occurs either through competitive substitution or through the complementary merging of products and services, or both at once (Dowling, Leichner & Thielman, 1998).

Convergence is a phenomenon by which previously separated sectors end up being part and parcel of one big metamarket (Valdani, 1997), thus generating an effective merging of previously different sectors.

A terminology specification appears to be right and proper—a specification adopted by the theory of business economics which conceptually defines the meaning of sector, industry, and market based on three dimensions of analysis.[1]

According to such terminology, the sector is characterised by all the customers' groups (first dimension), by all the required functions (second dimension), based on the same technology (third dimension). On the other hand, within the confines of such taxonomy, the concept of industry expresses the vertical relations among the sectors. Industry detects the globality of vertically linked sectors, starting from the production of raw materials up to the gleaning of products destined to intermediate or final consumption. Changes in the relations and among the confines of the different sectors are of the utmost importance in changing both the industry and the market structure.

Thus, based on the assumption that the definition of "sector" has to do with the uniformities evidenced on the offer side, the definition of industry relates to the process of productive verticalization, the definition of "metamarket" or complex market, widens the business's competitive area to include those products deemed as surrogate in terms of technology within those usage situations where similar functions and benefits are required.

The multimedia metamarket—generated by the progressive process of convergence involving the television, informatics, and telecommunication industries—comes to represent the "strategic field of action" where television operates in the digital era.

In its simplest form, convergence means the uniting of the functions of the computer, the telephone, and the television set. Taken literally, the unification of functions could produce a massive reorganization of a trillion dollars in global business (Yoffie, 1997).

The main issues in this process of convergence have been identified in the literature (Sculley, 1990; Bradley, Hausman & Nolan, 1993; Collins, Bane & Bradley, 1997; Yoffie, 1997; Valdani, 1997, 2000; Ancarani, 1999). Sculley in 1990 pictured the most well-known representation of sector convergence from the customer's perspective.

According to this perspective (discussed later by Yoffie in 1997), telecommunications, office equipment, consumer electronics, media, and computers were separate and distinct industries through the 1990s, offering different services with different methods of delivery. But as the computer became an "information appliance," businesses would move to take advantage of emerging digital technologies, such as CD-ROMs and virtual reality, and industry boundaries would blur (see Figure 1.1).

The field of action may be described through three main areas (see Figure 1.2)[2]:

- the strategic segments of potentially served users (end devices);
- the functions required by users (multimedia services);
- alternative technologies to meet those functions' requirements (multimedia networks).

Figure 1.1: The Convergence of Industrial and Service Sectors

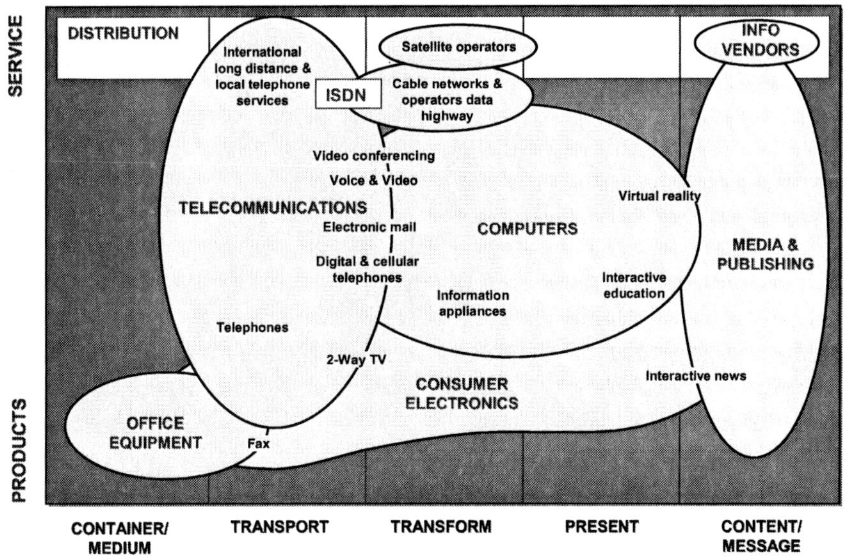

Source: Yoffie, Harvard University (1997)

We deem it appropriate to specify in as timely a manner as possible, that from the standpoint of terminology, the definition of multimedia service as used herein refers to a typology of service which incorporates more than one type of information (e.g., texts, audio, images, and video), transmitted through use of the same mechanism and supplying the user with the opportunity of interacting or changing the information.

However, the definition's key factors are represented by:

- opportunity of interactive use thanks to bi-directional transmission;
- simultaneous combination of different kinds of information (texts, images, data, video, and audio);
- digital transmission technology and signal compression.

The definition we have adopted does not limit in any way the concept of multimedia service to any particular transmission device such as the traditional telecommunication network, television, or Internet.

Figure 1.2: The Multimedia Market Dimensions

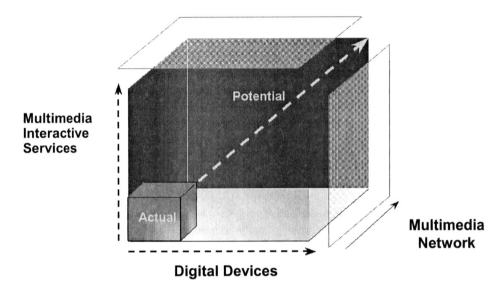

On the basis of the three areas, one can conceptually define the meaning of multimedia market where all the multimedia services destined to specific end devices are assembled, including all the technologies needed to meet those functions.

As a result of the convergence process, we cannot therefore talk about separate and different industries and sectors (telecommunications, television, informatics) since such sectors are propelled towards an actual merging of different technologies, supplied services, and the users' categories being reached.[3] A great ICT (Information Communication Technology) metamarket is thus originated.

THE KEY DRIVERS OF CONVERGENCE

In order to fully understand and describe the meaning of the process of digital convergence which characterizes and determines the multimedia metamarket, a survey of the main underlying drivers is in point.

We'll try to identify four drivers of industry convergence which have been identified in the literature (Bradley, Hausman & Nolan, 1993; Collins, Bane & Bradley, 1997; Yoffie, 1997, Valdani, 1997, 2000; Ancarani, 1999):

- at the macro level, we identify the legislation and the technological drivers;
- at the micro level, we identify the firm's strategy driver and the demand driver.

The *technology* constitutes the first relevant driver. It can be analysed according to three different characteristics:

- the digitalisation of signals and bandwidth;
- the increasing cheapness data production;
- the integration of transmission networks.

Digitalisation consists of the simple but enormous technological phenomena by means of which it is possible to translate data, sounds, and images into a single format, the digital binary format (i.e., transformed in a binary sequence, of numbers 1 and 0) or *byte*. The analog signal, through a series of successive procedures aimed at applying numerical algorithm,[4] is transformed into a digital signal unencumbered by any spatial or temporary redundancy.[5]

As a result, sounds, images, texts, and graphics are similar if not identical to each other once they have been translated into a digital signal. This means that they can be combined, stored, handled, and rapidly and efficiently transmitted over the same network and received by the same device.

One of the primary effects in the change of the nature and compression of the signal is the reduction in the use of the electromagnetic spectrum.

The advantages deriving from the change in signal transmission technology are plentiful, mainly as they relate to the typology of services being offered, allowing to use new hybrid products, characterized by integrated functions and which were previously usable only through separate product categories or allowing for the creation of new and additional functions.

As far as the second technological phenomenon is concerned, the computer shows an ever-increasing power, summarised in the well-known "Moore's Law,"[6] according to which the power and capacity of integrated circuits double every 18 months, while their cost remains the same.

Regarding the third phenomenon, the integration of transmission networks, the possibility of using a single transmission network for all signals means that there is a possibility of integrating previously separate infrastructures such as the telephone, television, and data, into a single "information highway."

These are known as multimedia networks, as it is not just possible to produce a message through a number of media (multi media), but users can also interact with contents in real time. A further technological explanation should be added to the above, which helps to understand the advent of the digital economy: Metcalf's Law. This law indicates that the value of a network is equal to the square number of its users. This means that the wider a network is, the more useful it is to each user. For a deep analysis of the technological drivers, see Wind and Mahajan (2000, 2002) and Hofacker (2000).

It's important to mention that the maximum impact has come from advances made in optical networking technologies (e.g., WDM) and TCP/IP Internet. The innovations made possible by Web/Internet have increased a recent push for converting and manipulating digital media, which has led to the increasing convergence of what used to be rival industry sector.

The *legislation drivers* refer to local, national, and international laws giving orientation to the ICT (Information Communication Technology) metamarket. Deregulation and liberalisation decisions (Bradley, Hausmann & Nolan, 1993) eliminate the monopoly conditions that are often present, opening the way to sudden changes in competition.

The degree of liberalization deeply influences the structure of the multimedia metamarket and the competitiveness level. A rapid deregulation is critical to keep competitiveness as well as being a *sine qua non* for a full spreading of multimedia applications.

At the micro level, however, there seem to be other drivers that can complete the explanation of industry convergence. In particular, it is worthwhile considering the drivers made up of *firm strategies (resource and knowledge based)* and those represented by convergent *demand needs (need clusters)*.

One of the driving forces of the convergence that is taking place is the *competitive dynamics* characterized by mergers, acquisitions, alliances, and other forms of cooperation—often made possible by deregulation—among operators at different levels of the multimedia value chain.[7] Competitive dynamics influences the structures of industries just as it does the typical managerial creativity of the single factory in originating products and services, combining know-how to create new solutions and removing the barriers among different users' segments.

A strategic intent is at play on the part of enterprises to use the leverage of their own resources within a framework of incremental strategic management in order to deploy them over an ever-increasing number of sectors.

One thinks, in this regard, of Hamel's (1996, 2000) concept of "driving convergence." This concept places the firm and its own competitive strategies at the control of the process of industry convergence.

In the multimedia market we witness the convergence of very different operators who contribute to the market with resources and know-how developed in their original fields of industrial action. What we are dealing with here is a know-how targeted at the transport of digitalized information, on the production of contents in the form of events—sports, movies, or likewise—on the ability to assemble such contents into an attractive offer, the ability to manage demand and production of additional assistance services.

Another driver in the process of convergence is the convergence in demand, in particular the formation of *integrated and convergent needs clusters*[8] (Ancarani, 1999). The expression "cluster of needs" means the tendency of customers to favour a single supplier for a set of related needs (Vicari, 1989; Busacca, 1994). The increasing technical complexity of many products leads customers to seek integrated offers that reduce the risks and costs related to the purchase of complementary products. If demand did not show need clusters that express the search for increasing integration of tools for the transmission of data, sounds, and images, the technological phenomenon of digitalisation would hardly be economically feasible, and the efforts of firms to lead the processes of industry convergence would be largely fruitless.

An exploratory model of industry convergence (Ancarani, 2001) is represented in Figure 1.3.

Figure 1.3: An Exploratory Model of Industry Convergence

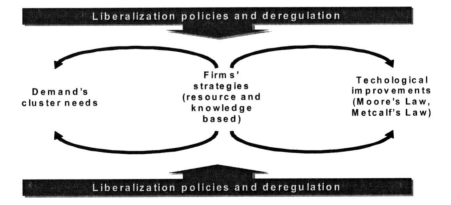

Source: Ancarani (2001)

THE NEW EMERGING MARKETS

Having conceptually defined the meaning of multimedia metamarket and highlighted the underlying key drivers, attention should be focused on the analysis of the new emerging markets as a result of the convergence process under way. A first systematisation of the convergence phenomenon, mainly focused on the technological and liberalisation drivers, is to be found in the works of Bradley, Hausman and Nolan[9] (1993) and Yoffie[10] (1997), and can be synthesized in the model put forth by Collins, Bane and Bradley[11] (1997). These authors present a first description of the new digital metamarket, emphasising the movement from a competitive environment characterised by vertical industries—sound communication (telephony), visual communication (television), and data (informatics)—towards a new multimedia industry, structured in five horizontal segments, where the players are trying to evolve new capabilities and establish new relations with the market partners.

These segments are:

- *Content production*, where the operating firms are mainly those producing books, cinema, music, television programmes, graphics, advertising, etc. This is a segment where producers of "intellectual works" operate.
- *Content packaging*, where the operating firms are those which do not produce the contents, but prepare them as product offerings in the end markets. They are publishers, cinema and television producers, and suppliers of information services.
- *Content transmission*, where the operating firms are those that supply infrastructure, in the first place telephone companies, television firms that use cable, and the suppliers of non-cable communications via aerial or satellite.
- *Content manipulation*, where the operating firms are typically involved in information technology. They supply the hardware, operating systems, and software that are needed to process information and images, allowing for the interaction between the sender and the receiver through the transmission network.
- *Content reception through devices*, where the operating firms are all the suppliers of equipment for the reception and reproduction of multimedia information—these are generally defined as information appliances.

These five horizontal segments form a new value chain in the converging digital industry. Within this new competitive environment, firms can supply a

Figure 1.4: The Model of Collins Bane and Bradley: Emerging Industry Structure

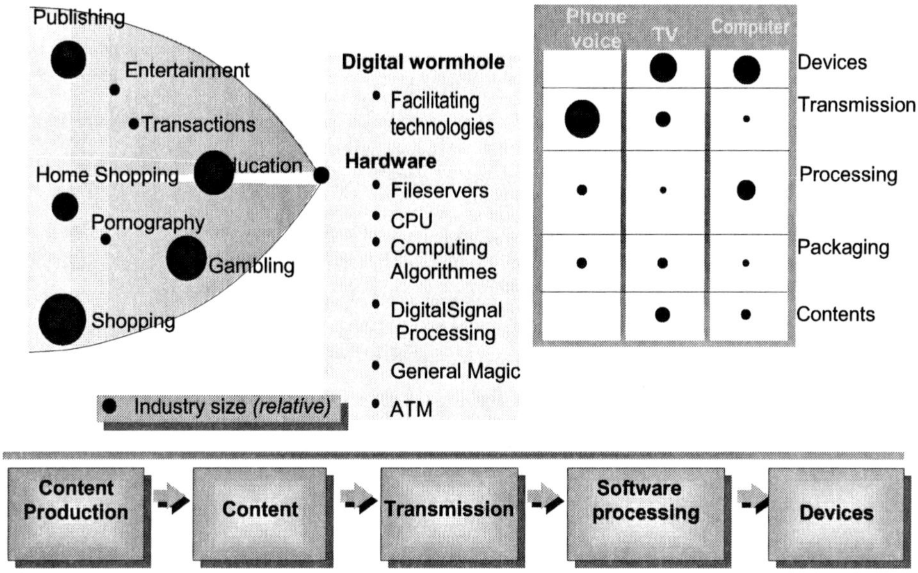

	Phone voice	TV	Computer	
Devices	●	●	●	
Transmission	●	●	·	
Processing	·	·	●	
Packaging	●	●	·	
Contents		●	●	

Publishing
Entertainment
Transactions
Home Shopping
Education
Pornography
Gambling
Shopping

● Industry size *(relative)*

Digital wormhole
- Facilitating technologies

Hardware
- Fileservers
- CPU
- Computing Algorithmes
- DigitalSignal Processing
- General Magic
- ATM

Content Production → Content → Transmission → Software processing → Devices

Source: Collins Bane and Bradley (1997)

homogeneous information product or service such as technologies and production, transmission, and reception tools, once they have acquired the information content from the respective producers (Ancarani, 2000). Further, firms that belong to the three traditional sectors have the chance to broaden their field of action. They can do this horizontally, through data communication, audio-visual and sound, and vertically, performing one or more of the stages of the multimedia value chain (see Figure 1.4).

The phenomenon has been further systematized by Wirtz in 1999 and—based on the results of such work—it is possible to identify the original core of the emerging digital metamarket in the process of technological and economic convergence among the informatics, telecommunications, media, and consumer electronics sectors. The convergence process presents effects which are amplifying in a chain-like fashion over other sectors such as entertainment, office equipment, commercial distribution, financial and insurance services, etc. The fundamental influence directions of such transformation processes are illustrated in Figure 1.5.

Figure 1.5: The Circular Convergence Model by Wirtz

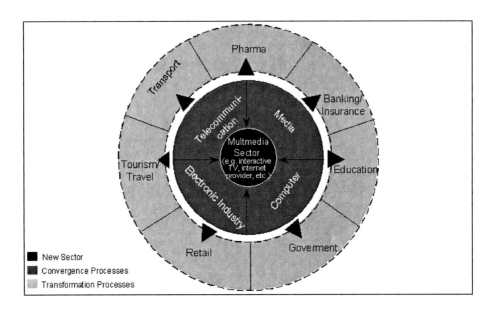

Source: Bernd W. Wirtz, Convergence Processes, *1999*

In the Wirtz model (1999), the different transformation processes under way are represented by three different circular rings (core ring, middle ring, and periphery ring). The convergence and transformation processes within the core ring originate new value chains and constellations of value which in turn lead to the creation of new markets (such as interactive television, which shall be analysed in Chapter 5). The emergence of the core ring will therefore have considerable cannibalisation effects on the essential business of the industries within the middle ring (media, computer, electronic industry, telecommunication). These four businesses tend to converge toward the central ring, giving rise to important transformations in the traditional production structures and constituting a new segment of multimedia market. The third peripheral ring detects the sectors undergoing transformation processes due to the process of convergence.

However, acknowledging the metamarket's dynamic nature does not in any shape or form either impinge on the contingent validity of its confines put forth herewith, nor does it stand in the way of drawing a timely description based on some homogeneity factors shared by businesses operating in this sector, according to an approach which has by now become standard procedure (Guerci, 1979; Abell, 1980; Day, 1984; Volpato, 1986; Sinatra, 1989).

We now focus our attention on the TV sector in the digital era in light of evolving directives characterizing the multimedia metamarket.

The highlighted convergence process forces us not to consider the TV sector as a separate sector any more, but rather as a sector strictly linked to the other converging sectors (telecommunications, electronics, computers).

First of all, the growing integration trend between personal computer and digital TV determines the birth of new emerging markets for interactive TV broadcasting and Web TV. As a consequence we see a progressive convergence of television contents distributed by diverging platforms (Personal Computer, TV set, 3G mobile phones). Changes in the user's personal behaviour and needs are a very significant driver in terms of the transformation process under way. The TV set can supply more and more interactive functions previously available only through the PC (Internet access, communication services, interactive services) while computers allow users to access functions once only TV related. This progressive convergence between informatics and TV medium creates relevant effects also on other sectors.

At the level of end device, the electronic entertainment industry is responding to the evolution process by expanding the traditional systems of TV reception (TV sets) and powering such systems with informatics know-how. It is indeed developing "smart" TV sets which carry out functions previously carried out by PCs.

A consequence of such a trend on a competitive level is that large electronic entertainment multinationals such as Sony, Philips, or Matsushita are carrying out specific acquisitions in the informatics hardware sector and in the interactive digital media (e.g., Matsushita holds a share of British Interactive Broadcasting in UK). Also the computer industry is increasing its traditional offer by supplying multimedia components and services.

Besides this transformation of the end devices' industry, TV media and telecommunications' carriers (via cable infrastructures and smart networks) are converging to the market segment of digital TV broadcasting which can offer entertainment, information, TV-commerce services (TV shopping and interactive advertising, TV banking and finance, games and betting) together with other interactive services through different technological manners of signal transmission[12] (see Table 1.1).

Copyright © 2003, Idea Group Inc. Copying or distributing in print or electronic forms without written permission of Idea Group Inc. is prohibited.

Table 1.1: Convergence Trends

Convergence trends	Selected example
Voice and data transport convergence	- Voice and data transport, e.g., on IP network using software developed by companies like Vocal Tec
Wireless/wireline convergence	- GSM/DECT launching
Communication and informatics hardware convergence	- Personal digital assistance such as Nokia 9000 communicator
ICT technology and content convergence	- Companies such as Matsushita and Sony entering the content business
Television and informatics convergence	- Interactive information transmitted via traditional TV - PCTV manufactured by companies like Compaq and Fujitsu
Convergence among industrial sectors	- Convergence among computer making, software development, and integrative functions systems in companies such as IBM
Convergence between TV and telephonics	- Telephonics offered via cable TV by companies such as C&WC in UK
Convergence of carrier services	- Wireline, wireless, Internet transmission, and other services in a single package

Source: Adapted from EITO, European Information Technology Observatory, 1998

THE ORGANIZATION OF THE NEW VALUE CHAIN

The value chain[13] of the new multimedia metamarket, generated by the aforementioned convergence process, can be subdivided into some phases, each one characterized by specific operators (see Figure 1.6):

- *content creation:* making of images, texts, music;
- *content production:* making of the final product by movie making, music, TV programs, graphics, advertising companies, etc. Makers of "works of intellect" operate in this segment;
- *multimedia packaging:* assembling of different contents and services (by TV channels, online service suppliers, press agencies) in order to create programs fit to be offered in the end or target markets;
- *service development and management:* offer of services of added value in terms of the created content such as EPG (*Electronic Programming Guide*), interactive services, customer services, etc;
- *content and service distribution:* supplying of the infrastructure needed to transmit the signal and manage its transmission for the target users (e.g., cable TV);

Figure 1.6: The Value Chain of the Emerging Multimedia Market

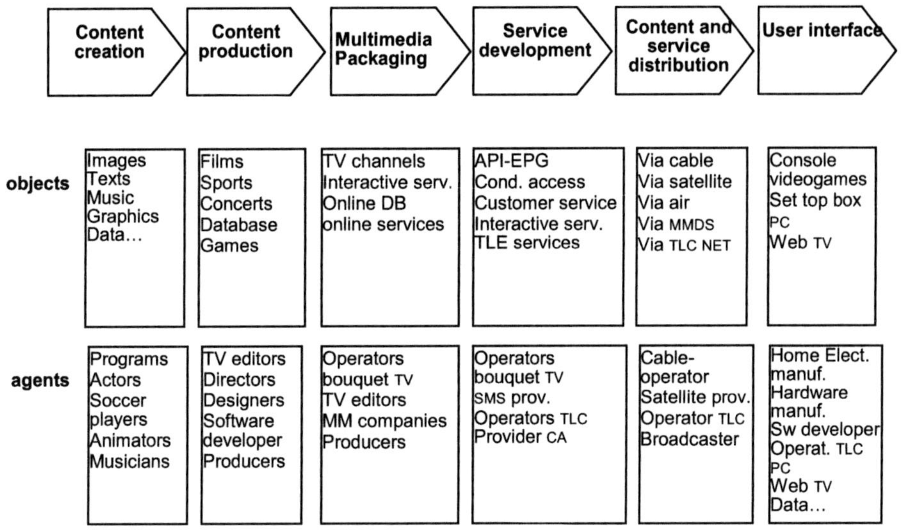

	Content creation	Content production	Multimedia Packaging	Service development	Content and service distribution	User interface
objects	Images Texts Music Graphics Data...	Films Sports Concerts Database Games	TV channels Interactive serv. Online DB online services	API-EPG Cond. access Customer service Interactive serv. TLE services	Via cable Via satellite Via air Via MMDS Via TLC NET	Console videogames Set top box PC Web TV
agents	Programs Actors Soccer players Animators Musicians	TV editors Directors Designers Software developer Producers	Operators bouquet TV TV editors MM companies Producers	Operators bouquet TV SMS prov. Operators TLC Provider CA	Cable-operator Satellite prov. Operator TLC Broadcaster	Home Elect. manuf. Hardware manuf. Sw developer Operat. TLC PC Web TV Data...

Source: Adapted from "L'industria Della Comunicazione in Italia," Istituto Economia dei Media, 1998

- *user interface*: production of end devices and manufacturing of the equipment required for reception and reproduction of multimedia information, generally defined as *information appliances*.

We may combine the different operators with the different phases of the value chain into four main categories, stressing the specific critical success factors (Figure 1.7).

As far as packagers are concerned a relevant factor is control of the content access in order to guarantee access to the most interesting and innovative contents (to this end it bears mentioning the many contractual agreements, primarily between TV channels and film production companies such as Flextech with Time Warner in the UK or Rai with Paramount in Italy).

A critical success factor for content providers and service providers is the independence which plays heavily on their bargaining power and which is determined by the size and attractiveness of the offered content.

Access suppliers—up to creating the required standards—are more and more influential in their interacting with packagers and distributors who are indeed progressively losing access control on the target user.

Figure 1.7: Critical Success Factors for Each Value Chain Phase

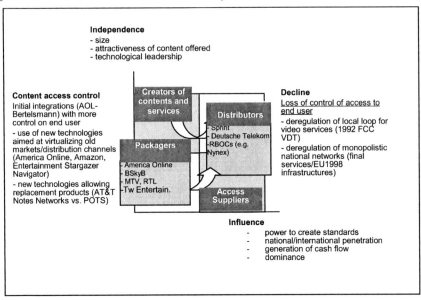

Figure 1.8: The TV Industry's Value Chain

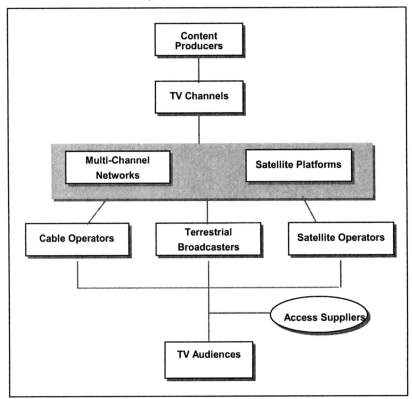

The New Value Chain of the Television Industry

After having defined the new configuration taken up by the value chain of the new multimedia metamarket, it seems appropriate to dwell further on how the value chain of the TV industry is changing (Figure 1.8).

We will now be analysing the strategic and competitive aspects featured in the single industrial segments which make up the TV industry's value chain in the multimedia metamarket, and we will be stressing the impact generated by the convergence process.

Content Providers

The role of content providers (film-making companies, TV producers, publishers) is the creation of TV programs and contents subsequently broadcast by the TV channels and by other digital devices (PC, PDA) through the different transmission technologies. Within this segment we can find companies focused on the creation of content (film-making companies) and companies integrated originally with the TV channels (such as Disney with the Disney Channel).

The increasing number of TV channels—following the TV signal digitalisation process—coupled with the possibility of conveying the digital content itself through different digital platforms, has generated a substantial increase in content demand. Content suppliers or broadcasters are the most critical area of the new emerging multimedia metamarket. The growing competitive pressure has led some television channels to choose a niche positioning in terms of offered contents (through theme channels or channels targeted to specific audiences). Therefore, quite a few deals have been made between content producers (such as the holders of rights for sport events and films) and digital TV broadcasters for Pay TV broadcasts. The critical success factors of content producers in order to acquire a strong bargaining power vis-à-vis the TV channels are as follows:

- size;
- attractiveness of the offered product;
- technological leadership.

TV Channels

The TV channels assemble the contents supplied by the content producers and assemble programs to be broadcast. This segment includes the traditional generalistic or theme channels, free access or encoded, analog or digital (e.g., CNN, NBC, Discovery Channel, Disney Channel, MTV etc.).

A critical success factor for the TV channel is the ability to access the best contents in terms of attractiveness, in order to conquer and create loyalty in the channel's target audience. Deals are struck with content suppliers for this target to be reached.

Multi-Channel Network

The multi-channel networks offer the TV market a package of channels which will be transmitted by satellite and terrestrial broadcasters or by cable operators. BSkyB stands as an example of a multi-channel supplier and offers a package including Sky One, Sky News, Sky Movies, and Sky Sports. DF1 in Germany offers CNBC, the Discovery Channel, and the NBC Super Channel. In Italy RAI offers the three analog terrestrial channels as well as other satellite theme channels.

Satellite Platforms

Satellite platforms broadcast the content of the channels making up their "bouquet" or package via satellite technology. The main European satellite broadcasters include BSkyB in UK, Canalsatellite and Canal Plus in France, and Tele+Digitale and Stream in Italy. Satellite platforms can broadcast the signal both in the free access and the encoded form, with transmission of the analog or digital signal both directly to the users who get the signal via their satellite dish, and to the cable operators who in turn, via the cable network, distribute the signal to the users.

Cable Operators

The cable companies are responsible for running the cable network, the hook-up to residences, and the transmission and reception of signals through the network. Often the cable networks' operations are limited to the supply of TV services.

In Europe, many of the bigger cable operators, e.g., Telia in Sweden and Deutsche Telekom in Germany, are also telecommunications operators since the required capabilities to operate in this sector are pretty much alike. These operators interact directly with the user and are responsible for collecting receipts.

Terrestrial Broadcasters

Terrestrial broadcasters transmit the traditional TV service via airwaves available in all the European countries.

Satellite Operators

Responsible for launching and operating/servicing broadcast satellites, such operators rent space on the satellite transponder to the satellite broadcaster. The main operators in Europe are Astra and Eutelsat.

ALLIANCES AT DIFFERENT LEVELS OF THE VALUE CHAIN

After having analysed how the multimedia value chain changes and which are the peculiarities of each phase as it pertains to the TV industry, we shall now deal with the modalities of evolution of the strategies adopted by companies operating in the different phases of the value chain (Figure 1.9).

From 1960 to 1980 companies have adopted strategies aimed at obtaining scale economies within mutual horizontal *core-businesses* (focalisation strategies). Although achieving scale economies (Chandler, 1990) could be a must for long-term success,[14] in time such condition proved to be insufficient. The growing convergence of different technologies and skills led to the realization

Figure 1.9: Value Chain and Main Competitive Strategies

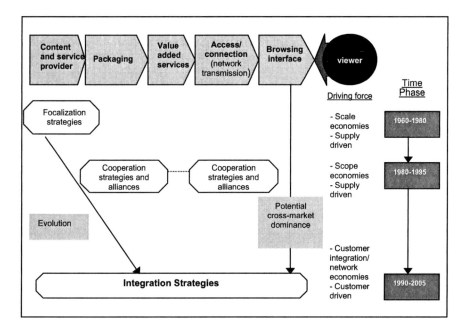

that companies with a wider scope could carve a profit for themselves which put them at an advantage over more strictly focalised companies. Companies expanded their activities to adjacent markets (1980-1985 phase) through cooperation strategies and alliances.

However, scale and target economies could not guarantee dominance in the converging industries. From 1995 to 2005 we are witnessing the development of integration strategies among the companies operating within the different phases of the value chain, aimed at obtaining network economies.

In order to overcome the user's resistance, the new converging technologies require a wide range of complementary products and services (online services, World Wide Web, hundreds of TV channels, and CD-ROMs with content-wise interest). Network economies furthermore allow for a cut in learning, distribution, and service costs.

Some interesting assessments emerge from an analysis of the different phases of the multimedia value chain and relations among the different operators (Figure 1.10).

Broadcasters, who have since time immemorial been developing primary skills in the phases of packaging and signal transmission (*network provider*), are developing many a contractual link with other players in the market.

There are several forms of "upstream" vertical integration with content suppliers (e.g., Flextech with Time Warner, RAI with Paramount) or with "downstream" access suppliers or cable network providers (Flextech with Telewest in the UK).

Many content providers are either integrating downstream in the packaging phase (e.g., Disney with the Disney Channel or News Corporation with BSkyB) or developing contractual links with network providers (Time Warner-AOL).

Figure 1.10 illustrates the location of the biggest players along the value chain and the relations among them.

Digital TV offers attractive opportunities to players who by tradition were foreign to the broadcasting world, particularly to companies operating in the field of telecommunications and informatics. Such companies are interested in exploiting digital television primarily as it pertains to interactive services and interconnection, software and hardware demands arising from them—either alone or in combination with traditional television contents. Different telecommunications (France Télécom, Telecom Italia, Telefònica, BT, AT&T) and informatics operators, first among them Microsoft, are now taking part in initiatives in the field of digital TV.

Figure 1.10: Location of the Main Players in the Value Chain and Relations Among Them

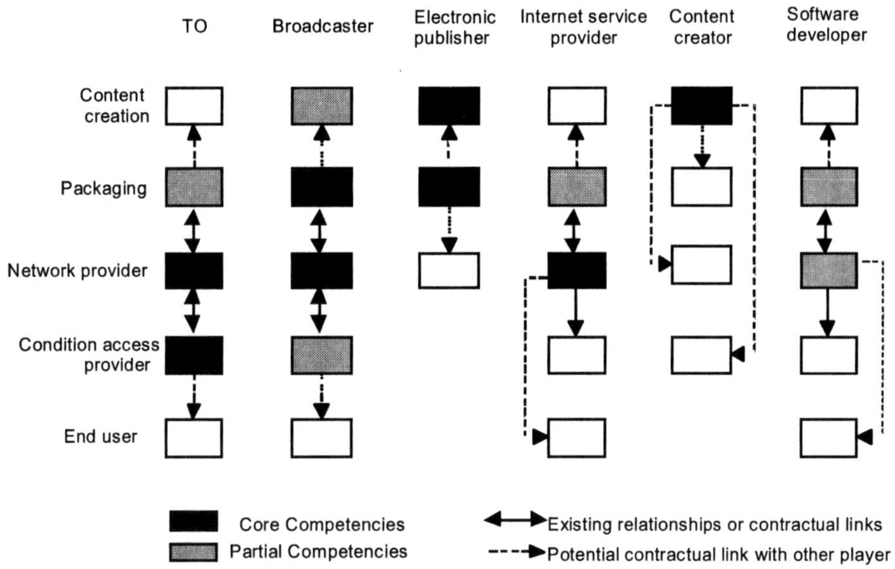

	TO	Broadcaster	Electronic publisher	Internet service provider	Content creator	Software developer
Content creation	□	▓	■	□	■	□
Packaging	▓	■	■	▓	□	▓
Network provider	■	■	□	■	□	▓
Condition access provider	■	▓	□	□	□	□
End user	□	□		□		□

■ Core Competencies ◄──► Existing relationships or contractual links
▓ Partial Competencies ----► Potential contractual link with other player

Source: Adapted from Squires, Senders Dempsey LLP and Analysys Ltd., 1998

Horizontal Integration Strategies

There are many examples of horizontal integration strategies which are achieved through alliances and mergers, both in the telecommunications sector and in the TV and multimedia content sector (Figures 1.11, 1.12, 1.13). The primary *driving forces* underlying the phenomenon are the following:

1. *Growth in the market power and scale economies achievement*
   The scale economies generated by the merger with the market leaders are conducive to an increase in profitability and competitiveness. In Europe, in cable TV and the Internet market, the following appear to be relevant factors:

   • the merger of Bell Cablemedia, NYNEX CableComms, and Videotron with Mercury Communications Ltd. which originated Cable and Wireless Communications;
   • the expansion of service provider Internet UUNet through acquisition of Unipalm PIPEX in UK and France.

2. *High cost of the new digital technologies*
   The substantial costs linked to the introduction of the new digital technologies drive companies to mergers, allowing an increase in the users' base and matching decrease of the contact cost (for example mergers and alliances in the media and distribution sectors or merger between Canal Plus and Nethold).

3. *Uncertainty in new services demand*
   The uncertainty in the demand is one of the main motives underlying alliances in the satellite digital TV sector. An important example is the one of Multimediabetriebsgesellschaft, which led Kirch, Bertelsmann, and BSkyB to the creation of a digital satellite platform in Germany.

4. *Internationalisation*
   The opportunity of increasing the market position in the offer of global services has led to agreements finalized to the supply of a critical mass in the ever-growing competitive system for international communication services.

5. *Opportunities deriving from regulatory changes*
   Many of the small cable TV operators carry out mergers and horizontal alliances in order to reach a progressive dimensional consolidation, allowing them to take an active role in the future generation of multimedia interactive services. Following the liberalization of the competitive context, thanks to the cable TV directive in Belgium—which allows cable TVs and other networks to supply telecommunication services—17 cable companies have joined US West operators in order to create a new

Table 1.2: Motives Underlying Horizontal Alliances and Mergers

Motive	Examples
Growth in market power/scale economies	Vebacom-Urbana Systemtechnik, Cable and Wireless Communications, AOL-CompuServe
High costs of new digital technologies	Canal Plus-Nethold
Uncertain demand for new services	Multimediabetriebsgesellschaft (Kirch, Bertelsmann, etc.)
Internationalisation	BT-MCI (WorldCom, GTE), Global One, UUNet-Unipalm Pipex
Opportunities deriving from regulation changes	MFSWorldCom, Telenet Flanders (17 Belgian cable TV companies and US West), NYNEX- Bell Atlantic

company, Telenet Flanders, with investment plans in new multimedia offers.

Figure 1.11: Examples of Horizontal Alliances and Mergers in the Telecommunication Sector

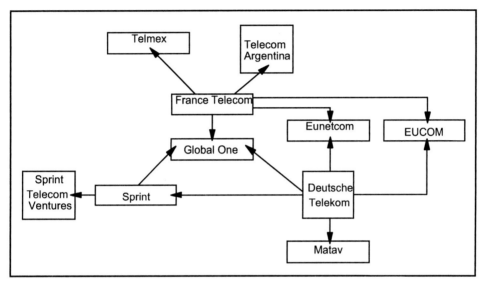

Source: Analysys 1998

Figure 1.12: Examples of Horizontal Alliances and Mergers in the TV Sector

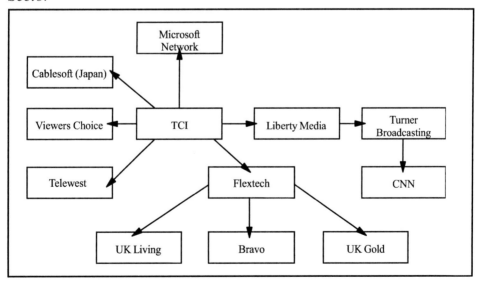

Source: Analysys 1998

Figure 1.13: Examples of Horizontal Alliances and Mergers in the Multimedia Content Sector

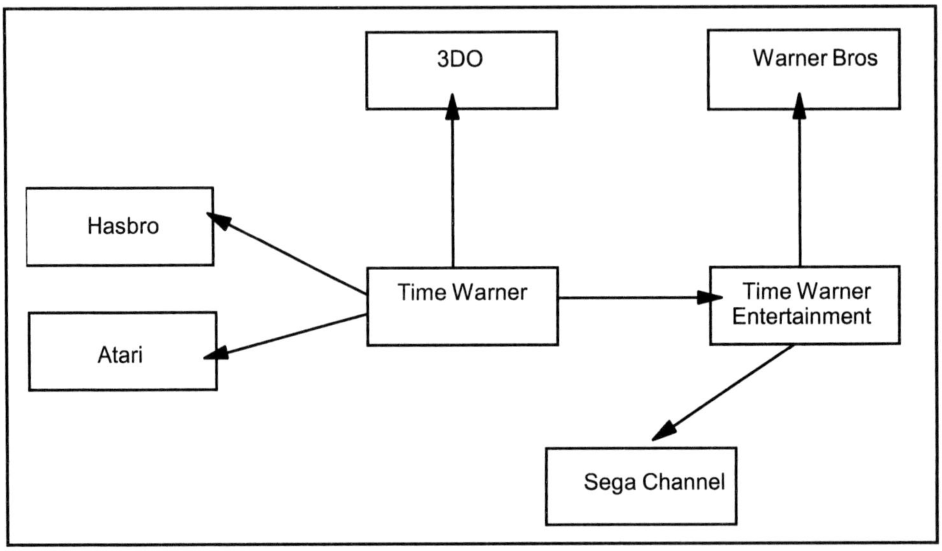

```
                    3DO                          Warner Bros
                     ↑                                ↑
                     |                                |
     Hasbro ←                                         |
              \                              → Time Warner
               Time Warner ─────────────────   Entertainment
              /                                        \
     Atari ←                                            ↓
                                                   Sega Channel
```

Source: Analysys 1998

Vertical Integration Strategies

Vertical alliances are determined by the need of the single company to develop along the vertical dimension of the value chain. The primary underlying *driving forces* are:

1. *Uncertainty in the market demand*
   Many companies have adopted vertical integration modalities to reduce the risks linked to the demand in new markets (for example the joint venture between Hughes Olivetti Telecom and DirecPC in 1996, set up to operate satellite communication and an Internet access service for residential customers and small businesses). The uncertain demand and the rapid market change have led, however, to short-lived alliances as shown by the case of TCI with Microsoft for development of Microsoft Network.

2. *Market positioning and access to new capabilities*
   Manners of acquiring the capabilities are required to establish a credible position in the market and positioning in multimedia market segments deemed strategically important. The alliance of Bertelsmann (publishing

and television activities) with America Online to form AOL Europe has enabled the group to increase network technical capability and supply Internet services. The BBC has developed Internet services and online information with ICL, which supplies the informatics capabilities required to run the service.

3. *User control possibilities*
   Many vertical mergers and alliances come into being in order to achieve more control of the user and increase the share of profits deriving from the sale of products and services. Disney Corporation has obtained access to many distribution channels by taking over ABC and Capital Cities in 1995, thereby accessing terrestrial and cable TV channels. BSkyB has developed new opportunities of distributing the new interactive services with BT (British Telecom) and potentially gaining access to the BT installation base. Similar advantages have driven Bertelsmann to cooperate with Deutsche Telekom in Germany.

4. *Shifting to value chain areas with a higher added value*
   This motivation is evident in many alliances formed among companies operating in the Internet market and in other phases of the value chain. Suppliers of mobile phone services such as Cellnet and Orange in the UK tend to shift in the supply of Internet services (with UUNET) in order to keep profitability within the framework of the communication market growth. Microsoft and NBC have created joint ventures to achieve synergies by exploiting the complementary abilities. The "new Internet channel" MSNBC enables NBC to add value to its basic TV activities through Internet and enables Microsoft (through Microsoft Network) to become a content supplier.

Table 1.3: Motives Underlying Vertical Alliances and Mergers

Motive	Examples
Precarious demand	Hughes Olivetti Telecom (DirecPC), @Home
Market positioning and access to new capabilities	Bertelsmann - AOL, BBC WorldWide - ICL, STET – IBM
Possibility of user channel control	BT - BSkyB, Disney – ABC - Capital Cities
Shifting to value chain areas with higher added value	Microsoft Network – NBC (MSNBC Internet news channel)
Facing competition of companies in connected sectors	US West - Time Warner, Oracle - Sun - Netscape (Network Computer)

Figure 1.14: Examples of Vertical Mergers and Alliances

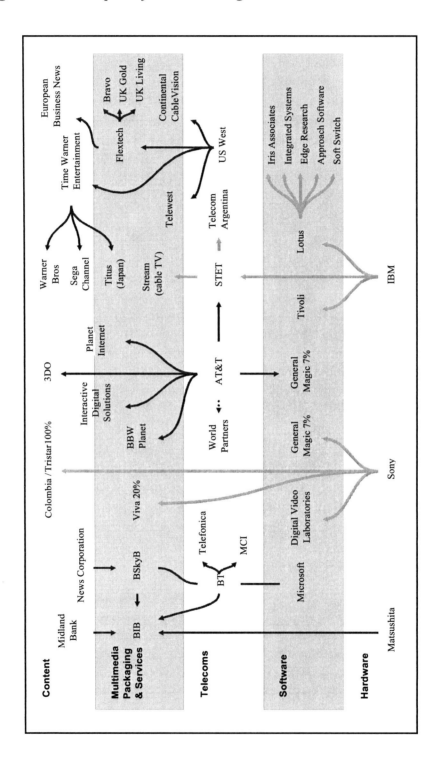

5. *Facing the competition of companies in connected markets*
   In order to face the competition of other companies in connected sectors, many alliances have been born, such as the investments by US RBOCs in cable TV aimed at preventing competition in cable companies entering the telecommunication market.

EVOLUTION OF THE OFFER

The process of technological convergence (Ames & Rosenberg, 1977; Abernathy & Uttenback, 1978; Bresnahan & Trajtenberg, 1995; Clark, 1995) among informatics, telecommunications, and television has increased the number of services offered by each of these industries' players.

From the analysis of the multiple dimensions, it is possible to define two basic forms of convergence that Greenstein and Khanna (1997) recognized as substitutes and complements.

1. The Substitute Paradigm (1+1=1)
2. The Cooperative Paradigm (1+1=3)

Two products can therefore converge into substitutes or into complement.[15] In the first instance we are dealing with a situation in which products are considered mutually interchangeable (on-demand video programming constitutes a very indicative example in this respect since it offers a replacing product compared to the use of videotapes), while in the second instance the products are better appreciated if they are grouped together and in any case much more than they were before the sectors converged (for example the integration of different interactive services with the audiovisual content).

Greenstein and Khanna[16] maintain that convergence can often start with complementary products, and subsequently such products may become replacements.

The replacement rate of the market may be highlighted by the analysis of four key variances (Figure 1.15):

- the *economic replacement* refers to the economic advantages generated by multimediality, compared to the present distribution system (e.g., the costs to access Video-On-Demand as opposed to the ticket price for a movie in a theater or the renting/purchase of a videotape);

- the *changes in the user's behavior* depend on the relative ease of use of alternative distribution mechanisms (for example the convenience of ordering a film through the phone in the case of Pay-Per-View rather than going to the nearest store to buy or rent a tape);
- the *technological evolution* raises the problem of the existence of a stable standard minimizing the consumer's investment risk;
- the *replaceability of the product* depends on the presence of infrastructure prerequisites.

However, it seems appropriate to point out that there is a form of duality shown by the fact that the creation of new capabilities is generally associated both to convergence forms in the replacements and in the complementary (Greenstein, 1999), albeit at different levels.[17]

As it has been pointed out at the beginning of the chapter, by multimedia service we mean a typology of services incorporating more than one type of information (e.g., texts, audio, images, and video) transmitted by using the same mechanism or device and enabling the user to interact or modify the information.

Figure 1.15: Factors Determining Products' Replaceability

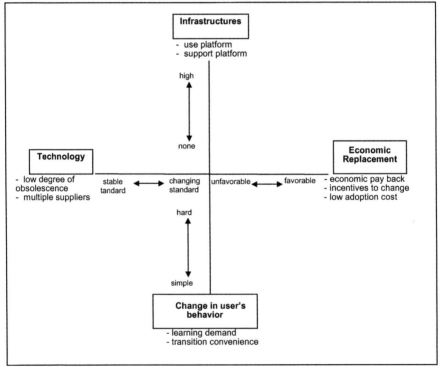

Source: Adapted from Mercer Management Consulting

Figure 1.16: Replacement Possibilities

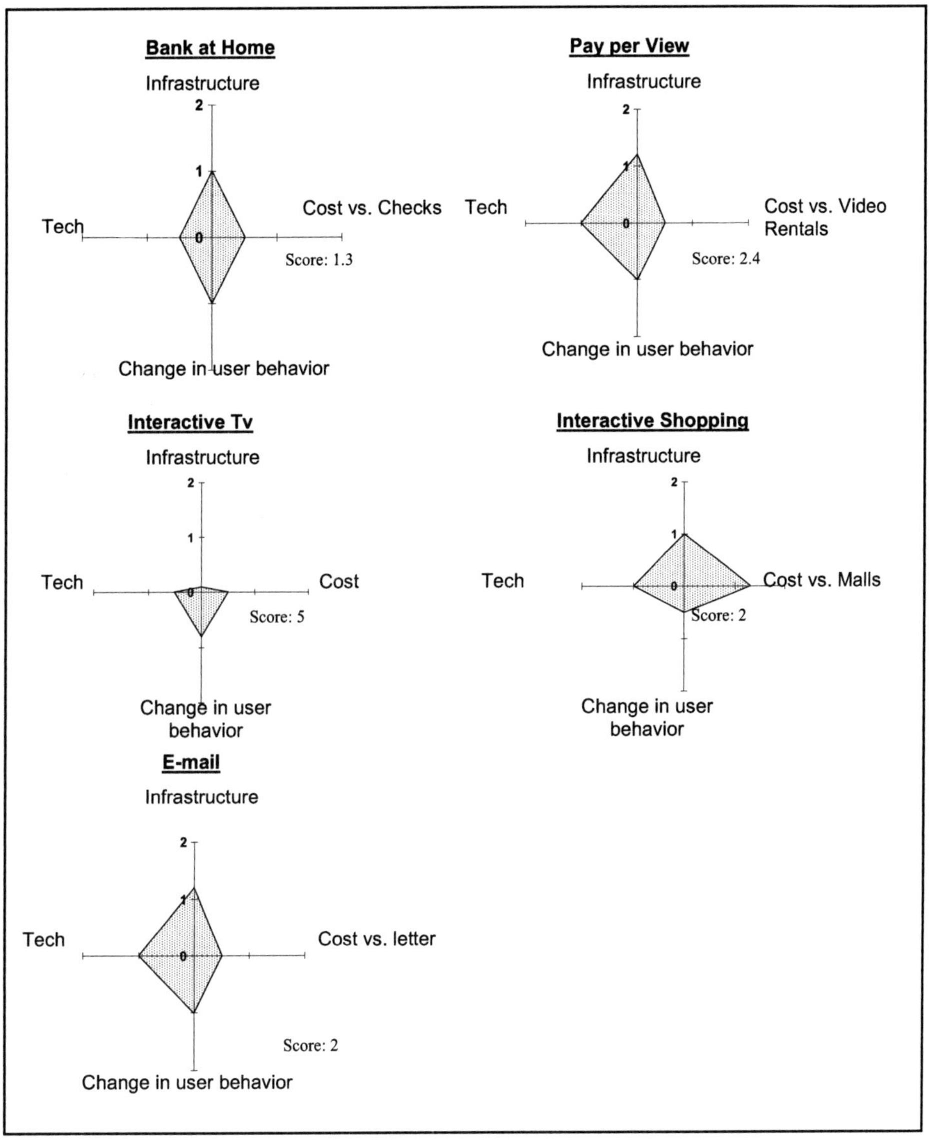

Source: Adapted from Mercer Management Consulting

Figure 1.16 takes into consideration five possible multimedia applications (home banking services, Pay-Per-View, interactive television, interactive shopping, e-mail), analysed on the basis of the four previously discussed areas, in order to assess their possible conversion to multimedia. The results show that the emerging multimedia services are e-mail, film on-demand, and interactive shopping.

The rapid increase rate of these markets will make the future replacement of other products easier by improving the technological infrastructure with subsequent changes in consumer behavior.

Summing up, the emerging multimedia services are generated by the convergence of new and old technologies, production methods, and marketing, and they pave the way to new business areas, employing at the same time existing skills and capabilities.[18]

In this respect, the "takeoff" of digital TV can be interpreted both as the first step towards the birth of multimedia communication systems and as the first concrete result of the progressive convergence process that is involving on a world scale the TV and telecommunications sectors. Such convergence process implies:

- at the *infrastructure level* the pooling of resources such as terrestrial, cable, and satellite networks for the transmission of TV signals;
- at the *service level* the introduction of new services presenting characteristics which are common to the traditional services—such as TV communication and telecommunications—but with the added bonus of offering innovation in the manner in which said traditional services are made available.

The evolution towards an ever-increasing integration between the two sectors is shaping a substantial change in the role and the functions traditionally carried out by TV and telecommunications operators who are more and more able to supply the same or similar services in a competitive fashion, developing

Figure 1.17: Direction of Innovations in Internet-TV Convergence

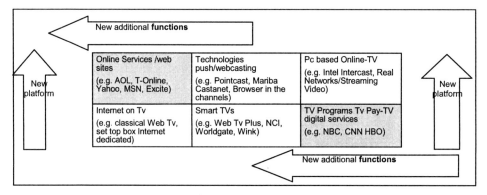

Source: Adapted from Thielmann & Dowling, 1999

forms of vertical (between running activities and service distribution activities) and horizontal integration (in which different service typologies are offered by the same operator or the same service is offered by non-traditional operators).[19] The sector of digital TV is characterized also by a progressive convergence with informatics (Web TV), allowing new, additional functions and the use of new platforms (Figure 1.17).

ENDNOTES

[1] Valdani, E. (1996) *Marketing Strategico—Un'Impresa Pro-Attiva per Sviluppare Capacità Market Driving e Valore*. Milan, Italy: ETAS Libri, p. 301.

[2] According to the theoretical conceptualisation put forth by Abell (1980), a business may be defined based on three dimensions: the groups of clients (i.e., "who" is the recipient of the service); the functions operated for the clients ("what" the clients themselves require); the technologies used ("how" the clients' needs are met). In reference to each of these three areas, business is further defined by the scope and differentiation of the very same areas.

[3] On this topic, please see Bartholomew, M.F. (1997). *Successful Business Strategies Using Telecommunications Services*. Noston, London: Artech House.

[4] A complex signal is applied to the TV signal by means of a Fourier transformation which can "map" a type of image onto another type, thus carrying out a complex mapping which can concentrate all the information around a more restricted area. This first movement—called algorithm—reduces spatial redundancy. Source: Vannucchi G. (1998). *La Storia Della Televisione Digitale*, http://www.mediamente.rai.it.

[5] Reduction of a temporal nature can be assimilated to the concept of compression: in the case of two television images, the comparison is made and the product being sent is not all the information, but only the information that is different between the two sets. Source: Vannucchi G. (1998). *La Storia Della Televisione Digitale*, http://www.mediamente.rai.it.

[6] According to Moore's Law the quantity of microelectronic processing speed, power, or memory that can be purchased with a dollar doubles every two years or so. In contrast, Internet traffic doubles two to five times per year. Named for Gordon Moore of Intel. Source: Owen, B. (1999). *The Internet Challenge to Television*. Cambridge, MA: Harvard University Press, p. 347.

[7] Collins, Bane, and Bradley have reported 508 multimedia alliances up until 1993. Since then activity has been increasing. In 1994, for example, according to KMPG

Peat Marwick, $27.8 billion have been spent on multimedia acquisitions and a further $22 billion in the first half of 1995. Source: Multimedia no-man's land. *The Economist,* July 22, 1995, p.57.

[8] Ancarani F. (1999). *Concorrenza e Analisi Competitiva.* Milan, Italy: EGEA.

[9] Cf. Bradley, S., Hausman, J. & Nolan R. (1993). *Globalization, Technology and Competition. The Fusion of Computers and Telecommunications in the 1990s.* Boston, MA: Harvard Business School Press.

[10] Yoffie, D.B. (1997). *Competing in the Age of Digital Convergence.* Boston, MA: Harvard Business School Press.

[11] Collins, D.J., Bane, W.P. & Bradley, S.P. (1997). Industry structure in the converging world of telecommunications, computing and entertainment. In Yoffie, D.B. (Ed.), *Competing in the Age of Digital Convergence.* Boston, MA: Harvard Business School Press, pp. 159-201.

[12] On this topic, please see Abe, G. (1997). *Residential Broadband.* Indianapolis, IN: Cisco Press.

[13] This conceptual model refers to a company as a complex of activities, each of which produces a value for the final or end user. Porter, M. (1985). *Competitive Advantage: Creating and Sustaining Superior Performance.* New York: The Free Press.

[14] Alfred Chandler (1990) in *Scale and Scope* op. cit. maintains that scale and target economies are able to achieve lasting competitive advantages for future generations.

[15] Greenstein, S. & Khanna, T. (1997). What does convergence mean? In Yoffie, D.B. (Ed.), *Competing in the Age of Digital Convergence.* Boston, MA: Harvard Business School, Free Press, pp. 203-207.

[16] Greenstein, S. & Khanna, T. (1997), What does convergence mean? In Yoffie, D.B. (Ed.), *Competing in the Age of Digital Convergence,* Boston, MA: Harvard Business School, Free Press, p.204.

[17] Greenstein, S.M. (1999). Industrial Convergence. In Dorf, R. (Ed.), *The Technology Management Handbook,* Boca Raton, FL: CRC Press.

[18] Yoffie, D. (1997). Introduction chess and competing in the age of digital convergence. In Yoffie, D.B. (Ed.), *Competing in the Age of Digital Convergence.* Boston, MA: Harvard Business School, Free Press, p.15.

[19] On this topic, please see European Commission. (1997). *Green Paper on the Convergence of the Telecommunications, Media and Information Sectors, and the Implications for Regulation. Towards an Information Society Approach.* Bruxelles, adopted by the EU Commission on December 3, 1997.

Chapter II

New Digital Media and Devices: Measuring the Potential for IT Convergence at Macro Level*

INTRODUCTION

As discussed in the previous chapter, the technological innovation process has a pervasive influence on the whole digital metamarket featured by the gradual convergence of three traditionally distinct sectors: IT, telecommunications, and media (Sculley, 1990; Bradley, Hausman & Nolan, 1993; Collins, Bane & Bradley, 1997; Yoffie, 1997; Valdani, 1997, 2000; Ancarani, 1999; Pagani 2000). The numerous innovations that could lead to "convergence" between TV and online services occur in various dimensions (Figure 2.1).

The *technology dimension* refers to the diffusion of technological innovations into various industries. The growing integration of functions into

---

* Earlier version of this paper was presented at the IRMA (International Resource Management Association) Conference in Seattle, May 2002. The author wishes to acknowledge the input of participants at this conference.

Figure 2.1: Dimension and Basic Forms of Convergence

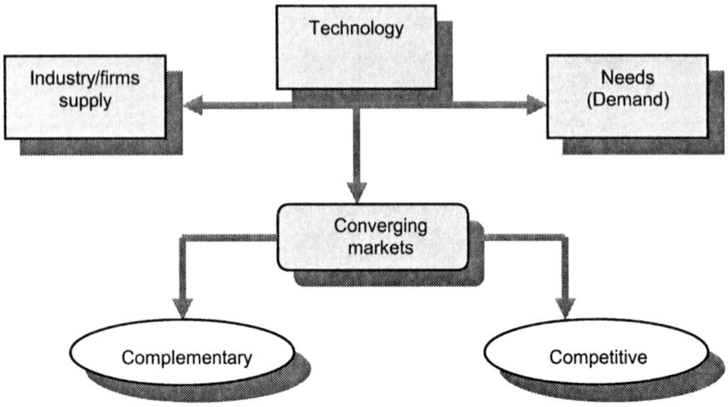

[Figure: A diagram showing "Technology" at the top, with arrows connecting to "Industry/firms supply" on the left and "Needs (Demand)" on the right, all feeding down into "Converging markets", which branches into "Complementary" and "Competitive".]

Source: Adapted from Dowling, Lechner & Thielmann, 2000

formerly separate products or services, or the emergence of hybrid products with new functions, is enabled primarily through digitalisation and data compression. Customers and media companies are confronted with technology-driven innovations in the area of transport media as well as new devices.

Typical characteristics of these technologies are digital storage and transmission of content from a technical perspective and a higher degree of interactivity from the user's perspective (Schreiber, 1997).

The *needs dimension* refers to the functional basis of convergence: functions fulfill needs of customers which can also merge and develop from different areas. This depends on the customers' willingness to accept new forms of need fulfillment or new products to fulfill old needs. When effective buying power creates a significant market demand for integrated functions, then boundaries are likely to be dissolved between different consumer groups (Grant & Shamp, 1997).

The *industry and firm dimension* refers to relevant industry variables that affect convergence.[1] Market barriers to convergence include industry cultures and traditions, regulation and antitrust-legislation prohibiting the creation of alliances, mergers & acquisitions. Deregulation often leads to a removal of artificial barriers that then promotes industry convergence. Firm-specific barriers to convergence include differences in company cultures and core competencies. Different activities along or across traditionally separated value chains may be merged by "management creativity" (Yoffie, 1997) such as the creation of new businesses, acquisition, or the creation of strategic alliances and networks.

Convergence describes a process change in industry structures that combines markets through technological and economic dimensions to meet merging consumer needs. It occurs either through competitive substitution or through the complementary merging of products or services, or both at once (Greenstein & Khanna, 1997).

The problem is that the notion of "convergence" itself is generally taken to be a characteristic of digital media, suggesting a possible future in which there might just be one type of content distributed across one kind of network to one type of device. Convergence remains ill defined particularly in terms of what it might mean for businesses wishing to develop a new media strategy.

This chapter argues for a definition of convergence based on penetration of digital platforms and the potential for cross-platform Customer Relationship Management (CRM) strategies, before going on to develop a convergence index according to which different territories can be compared. The model herewith discussed specifically refers to the European competition environment.

CONVERGENCE DISCUSSED

In general, the concept of digital convergence is used to refer to three possible axes of alignment: convergence of devices, convergence of networks, and convergence of content (Flynn, 2000). Although there is evidence in digital environments of limited alignment in some of these areas, there are considerable physical, technical, and consumer barriers in all three areas. Rather than convergence, the transition from analogue to digital is often being accompanied by a process of fragmentation.

This paragraph first shows a classification of new digital media and devices, and then describes what the limits to convergence are, understood as a phenomenon that involves the bringing together, merging, or hybridisation of different types of digital device, network, or content.

A Classification of New Digital Media and Devices

Based on a simplified version of the general communication model[2] (Shannon & Weaver, 1949), it is first possible to see how the communication flow between Media Company (source) and user (destination) is made possible by two interrelated elements: Transport Media (Transmitter or Channel), and End Device (Receiver).

Copyright © 2003, Idea Group Inc. Copying or distributing in print or electronic forms without written permission of Idea Group Inc. is prohibited.

Figure 2.2 : Transport Media and Reproduction Devices

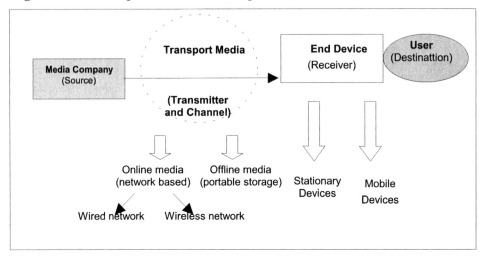

Source: *Adapted from Rawolle & Hess, 2000*

Transport media technologies belong to one of two categories: online (i.e., network based) and offline (i.e., portable storage)[3] media. Online media, also defined as network based, include wired network and wireless network. End Devices are subdivided into stationary and mobile devices (Figure 2.2).

Transport Media

Transport Media can be divided into two main categories: online media (i.e., network based) and offline media (i.e., portable storage).

Online media—also defined *network based*—include:

- *wired network:* all kinds of public networks are usually based on a backbone infrastructure and different access technologies ("last miles"). For the media industry especially, the Internet and TV cable are relevant wired networks, because both can be used for the distribution of digital contents[4];
- *wireless network* allowed the transmission of contents without having a physical link. Traditionally, the media industry used terrestrial broadcast or satellites, which are both constrained to simple transmission from sender to receiver. Mobile communication technologies, like the Global System Mobile Communication (GSM), support two-way transmission of data and can be used for Internet access via Wireless Application

Protocol (WAP), but provide significantly lower bandwidth at present. New technologies, like the General Radio Packet Service (GRPS) or the Universal Mobile Telecommunication System (UMTS), will improve bandwidth dramatically.

Offline Media (i.e., portable storage), like Compact Discs (CDs) or Digital Versatile Discs (DVDs), can be used to distribute digital contents through traditional retailers.

End Devices

End devices can be divided into stationary and mobile devices (Rawolle & Hess, 2000, p.89). Desktop Personal Computer (Desktop PCs) and television sets (TV sets) can be considered the most important examples of stationary devices. Besides technical differences (less information processing and storage capabilities, less interactivity, higher quality for replaying video and audio contents), the main distinction between PCs and TV sets is user behaviour.

Whereas PCs are typically used for information-based purposes (retrieval and processing) in an interactive way, the TV is usually utilised in a more passive and also entertainment-oriented manner.[5]

At present, consumers spend a greater share of their daily media budget on watching TV than using the PC-based Internet (ARD, 1999, pp.68-85).

Mobile Devices can be divided into multipurpose and special purpose. Multipurpose equipment like notebooks or subnotebooks have similar reproduction and processing capabilities, similar to desktop PCs. Personal Digital Assistant (PDAs) are considerably less powerful than notebooks.

Next to multipurpose devices there are other specialised appliances like eBooks for reading, mp3-players for music, and a set of Internet appliances that focus on Web access and mobile phones for speech-oriented interpersonal communication.

Mobile phones are increasingly used for Internet access via Wireless Application Protocol (WAP), and they represent a new target device for the media industry. Mobile phones and PDAs will probably merge into one device (see Luxa 1999, p.173). Some of these products might even support mp3 for audio replay.

The Matrix of Transport Media and End Devices

From the technologies described above, numerous combinations of transport media and end devices may be constructed. The nine-field matrix adopted

Figure 2.3: Relevant Categories of Transport Media/Device Combinations

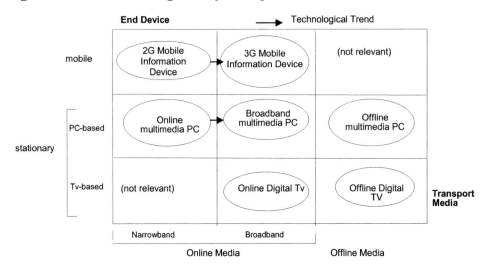

Source: Adapted from J. Rawolle, T. Hess, 2000

by Rawolle and Hess in 2000 summarizes relevant categories of transport media and devices combinations[6] (Figure 2.3).

As mentioned above, end devices can be divided into stationary (TV-based vs. PC-based) and mobile.

With regard to Transport Media, a difference must be made between online and offline media.

The resulting categories (as shown in Figure 2.3) are:

- *Mobile Information Device (MID):* This category pools available mobile devices with online capabilities. Typical examples of MIDs are Wireless Application Protocol (WAP) compliant mobile phones and PDAs that access the Internet over low bandwidth mobile communication networks. Usually, these appliances are restricted in terms of processing power, memory, display capabilities, and input facilities.
- *3G Mobile Information Device (MID 3G):* Technological advances are leading to substantial improvements in both mobile transmission technology and end device capabilities.
- *Online Multimedia PC:* One of the most popular platforms to access the Internet are stationary desktop PCs coupled with ordinary modems or ISDN connections.

- *Broadband Multimedia PC:* They offer the possibility to access the Internet with a substantially higher bandwidth. An additional technical option is to receive television via TV-cable.
- *Offline Multimedia PC:* Desktop PCs with CD-ROM or DVD drives.
- *Online Digital TV:* Their main focus is to receive broadcast-oriented contents, although feedback channels are possible from a technical standpoint.
- *Offline Digital TV:* As well as PC-based end devices, television sets support being connected with DVD drives. DVDs offer significant advantages over analogue videotapes and are therefore expected to replace the latter within the next few years (Sedman, 1998).

Barriers to Convergence

There are, in principle, two different types of constraints on the convergence of the devices that are used to access the three digital platforms (digital TV, PC/Internet, and mobile telephony). These two constraints—technical constraints and consumer-based ones—can be summed up in the following three questions:

1. Is it physically possible to merge the two devices?
2. Is it technically possible to merge the two devices?
3. Will consumers want to use the merged device?

Clearly, even where the answer to questions 1 and 2 is "yes," if there is no consumer adoption of the resulting hybrid, convergence will not take place.

Given that we are talking about three different types of network access devices here (the TV, the PC, and the mobile phone), there are three potential areas of convergence—PC/TV, PC/mobile phone, and TV/mobile phone.

In the *physical domain* the barriers to PC/TV convergence lie principally with respect to the size of the input device and its portability. The barriers to PC/mobile phone convergence in the physical domain are rather more acute and there is a divergence along every physical measure (size of display device, size of input device, and portability).

A mapping of the relevant physical characteristics of different types of consumer device is shown in Table 2.1.

Technical requirements either affect the available transport media, the addressed end device, or both. Three important aspects dominate in this area:

Table 2.1: Physical Characteristics of Consumer Devices

Characteristic	TV	PC	Mobile phone
Size of display device	Large	Large	Small
Size of input device	Small	Large (keyboard)	Small (keypad)
Portability	Low	Medium	High

the access mechanism, the number of simultaneous recipients, and the support of feedback channels in case of transmission media.

With regard to the access mechanism, a distinction between push and pull mechanism must be made. Pull-oriented access is characterised by the data transmission being triggered by the end user (which is typical for Web applications or Video-On-Demand), whereas push-oriented transmission is triggered by the sender. Push services can be time scheduled (e.g., television broadcast). Furthermore, push services can address one or more recipients[7]; respectively, a distinction between broadcast-oriented (e.g., television, radio) and unicast-oriented services (e.g., Web applications) has to be made (Kauffels, 1994).

Device-specific requirements mainly affect reproduction, storage, capabilities, and input facilities. Displaying and synchronising different kinds of media types is a basic demand with regard to reproduction. A distinction between static (time-invariant as text, graphics, and pictures) and dynamic (time-variant as video and audio) media types has to be made (Grauer & Merten, 1996).

Next, storage capabilities enable synchronous download and consumption of contents in case of online-media usage. Typically, end devices with roots in information technology (like PCs, PDAs, and notebooks) possess sufficient persistent storage capacity. In contrast, most of the entertainment electronics lack comparable characteristics.

Another important aspect of end devices are input facilities. Typically PC-based end devices possess the most advanced mechanisms for user input (keyboard, mouse, joystick, etc.). In contrast, mobile or TV-based devices usually lack sophisticated input facilities.

A comparison among the relevant *technical characteristics* of the three different types of consumer device (Table 2.2) shows that there is little evidence of TV/mobile phone convergence as yet, and in any case the technical constraints with respect to this particular combination are implicit in the consideration of the other instances.

Table 2.2: Technical Characteristics of Consumer Devices

Characteristic	TV	PC	Mobile phone
Display type	Cathode ray tube	Cathode ray tube	Liquid crystal display
Display resolution	Medium	High	Low
Display scanning mode	Interlaced	Progressive	Progressive
Display refresh rate	Medium	High	High
Processing power	Low	High	Low
Storage	Low	High	Low
Power requirement	High	High	Low

Consumer attitudes[8] towards devices that inhabit the (analog) TV environment, as opposed to the (digital) PC and (digital) mobile telephony ones, are also widely different. End users have certain usage patterns and behaviours that are closely correlated to end devices and transport media. As mentioned above, PC usage differs from TV usage in terms of user activity (active vs. passive) and purpose (information and entertainment). Another important aspect has to be considered in view of user attention. Table 2.3 serves to differentiate some relevant characteristics.

Table 2.3: Differing Consumer Expectations for Different Platforms

Consumer expectations in Tv space	Consumer expectations in PC space	Consumer expectations in the mobile phone space
Medium, stable pricing of goods	High, unstable pricing of goods	Low unstable pricing of goods
Infrequent purchase (once every 7-11 years)	Frequent purchase (every 18 months to 3 years)	Frequent purchase (every 18 months to 3 years)
Little requirement for software and peripheral upgrades	High requirement for software and peripheral upgrades	Medium requirement for software and peripheral upgrades
Works perfectly first time	Probably will not work perfectly first time	Probably will work first time
No boot-up time	Long boot-up time	No boot-up time
Low maintenance	High maintenance	Low maintenance
Low user intervention	High user intervention	High user intervention
Little or no technical support required	Substantial technical support required	Little technical support required

Table 2.4: Different Content Characteristics of the Three Major Digital Platforms

Tv/broadcast content attributes	PC/Internet content attributes	Mobile telephony content attributes
Video heavy (moving pictures lie at its core, rather than text)	Video light (text and graphics lie at its core, rather than video)	Voice based (audio lies at its core, rather than text, graphics or video)
Information medium (the factual information transmitted is not vey dense)	Information heavy (the factual information transmitted is dense)	Where non voice based material is transmitted, it is information light (any textual information transmitted in an SMS message or on a WAP phone is sparse)
Entertainment based (to provide a leisure activity rather than learning environment)	Work based (to provide work related or educational information or to enhance productivity rather than to be entertained)	Both work based and socially based (to provide work related information or to enhance productivity rather than to be entertained).
Designed for social o family access	Designed to be accessed by solitary individuals	Designed to be accessed by two individuals
Centrally generated (by the service provider)	Both centrally generated (content on a CD-Rom or website) and user generated (email, chat, personalisation, etc)	Predominantly user generated
User unable to influence content flow which is passively received rather than interacted with and linear in form	User tipically interacts with the content producing a non linear experience	Where centrally generated content is provided, user tipically interacts with the content, producing a non linear experience
Long form (the typical programme unit is 25 minutes long)	Short form (video information tends to be in the form of clips or excerpts).	Short form (text and websites highly abbreviated, audio in form of clips or excerpts).

The types of content that are carried over PC/Internet, broadcast, and telephony networks show some sharply differentiated characteristics and consumer usage, and distribution differs across platforms (Table 2.4).

The transition to all IP-based networks that are now carrying voice (VoIP) and video (cable TV with Internet capability) is leading the convergence in the network space.

There are many network differences (up/downstream capacity and interactivity) and authoring differences (development platforms, security, and standards), but consumers continue to use multi-platform content/commerce.

The ability to merge data about consumer preferences and transactional profiles across platforms is critical for any interactive media business, and this can be achieved through a process of cross-platform tracking. New killer applications are found by understanding the consumer and leveraging current consumer relationships.

The business potential in x-media commerce is in attracting, retaining, and capitalising on customer relationships through interactive media channels. This suggests a definition of convergence based not on the merging of digital devices, networks, or content, but on the extent to which the transition to two-way digital networks facilitates "consumer convergence" or cross-platform Customer Relationship Management[9] (CRM).

A DEFINITION BASED ON PLATFORM PENETRATION AND CRM POTENTIAL

The attractiveness or profitability of an industry is heavily determined by five competitive forces (Porter, 1980), including the rivalry between existing competitors. The model of analysis of converging industries with customer focus (Dowling, Lechner & Thielmann, 2000), shown in Figure 2.4, may help to structure the complexity by using a functional definition of the industry boundaries when analysing the threat of substitutes and likelihood of market entry. The focal point should be the customer and his/her needs.

Figure 2.4: Analysis of Converging Industries with Customer Focus

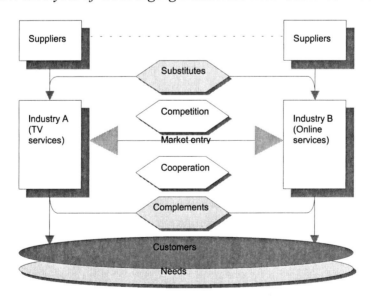

Source: Adapted from M. Dowling, C. Lechner, B. Thielmann, 2000

The previous paragraph has argued for a definition of convergence based not on the merging of digital devices, networks, or content, but on the extent to which the transition to two-way digital networks facilitates "consumer convergence" or cross-platform CRM. Customer Relationship Management can be described as the process of attracting, retaining, and capitalising on customers; it defines the space where the company interacts with the customer.

From a CRM perspective, the extent to which there is (or is not) technology-based convergence taking place between different platforms is not a central concern. The key is that these different, often incompatible, technology platforms enable customers to interact with companies through different channels, allowing those companies to increase the number of potential contact points with their customers.

At the heart of CRM lies the objective to deliver a consistently differentiated and personalised customer experience, independently from the interaction channel.

For media companies looking at their digital investment strategy in a given territory and seeking to maximise their benefit from this type of convergence, it is key to know which territories exhibit the best potential for development, so that those companies can decide where to initially test and/or introduce interactive applications or how to assess the likely success of existing projects in an eCRM context. The goal of the following model is to provide a methodology for convergence measurement. This model was applied to European countries and offers an interesting comparison frame among convergence levels in the different countries under scrutiny.

The following three indicators for the measurement of convergence potential are considered:

- critical digital mass index
- convergence factor
- interactivity factor

The Concept of "Critical Digital Mass Index"

One cornerstone in the measurement of convergence potential is the extent to which digital platforms (such as digital TV, PC/Internet access, and mobile telephony) are present in a given territory. This will obviously make it easier to reap the efficiencies and economies of scale that eCRM offers.

However, since eCRM strategies derive their greatest benefits across multiple channels, one needs to measure the penetration of such platforms in

Figure 2.5: Critical Digital Mass in Europe 2001

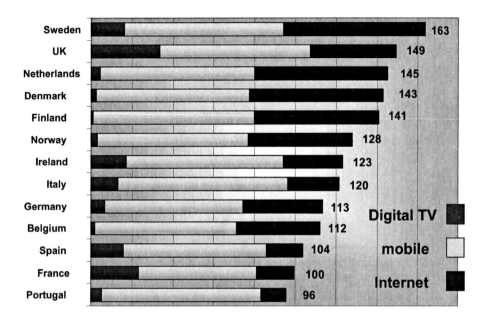

Source: New Media&TV-Lab, Bocconi University, 2002

combination. This combined measure (penetration of platform A, plus penetration of platform B, plus penetration of platform C) indicates the "critical digital mass" of consumers in any given territory.

The *critical digital mass* index for a territory is created by adding together the digital TV penetration, mobile telephony penetration, and PC Internet penetration in each territory from data at the end of 2001.

The Convergence Factor

The potential for eCRM is greatest where the same consumers are present across all three digital platforms: this would be the optimal situation for an integrated multi-channel eCRM strategy. The degree of overlap tends to be much higher when overall digital penetration is higher (this is not a linear relationship). If penetration of digital TV, PC/Internet access, and mobile telephony are all above 50%, the number of consumers present across all three platforms is likely to be much more than five times greater than is the case if penetration is at only around 10% in each case. Figure 2.6 illustrates this effect

Figure 2.6: Platform Overlap Increases Faster than Platform Increases

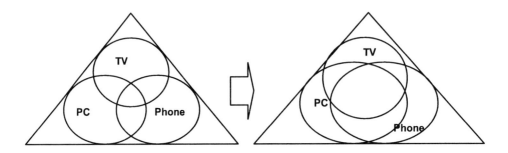

at work (the area within each triangle represents the boundaries of the total consumer universe).

This means that the critical digital mass indicator needs to be adjusted upwards for higher overall penetration levels.

The study uses simple probability theory to give a way of measuring the rate at which cross-platform populations increase as penetration of those platforms increases.

The "convergence factor" is derived from the penetrations of the three platforms multiplied by each other; the convergence factor for a territory is calculated according to the following formula:

(Penetration of digital TV/100) x (Penetration of mobile telephony/100) x

(Penetration of PC Internet/100) x 100

In order to give some explanation how the formula has been derived, we can suppose that the penetration of Platform A is 10%. What is the likelihood that one person chosen at random from the population is a member of that platform? Clearly, the chances are 10:1 against. If penetration of Platform B is also 10%, then the likelihood that the person we have chosen at random will also be a member of Platform B is 100:1 against.[10]

If penetration of Platform C is, again, 10%, then the odds that our initially chosen person is on that platform, too, is 1,000:1 against. In a population of one million individuals, in other words, the chances are that there are 1,000 people who fall into this category.

Take the opposite end of the penetration case, however, and assume that Platforms A, B, and C all have a penetration level of 90%. Using the same methodology, in a population of one million, the chances are that just over seven out of 10 people are on all three platforms (0.9 times 0.9 times 0.9); in a population of one million, this is equivalent to 729,000 people. Finally, of course, when all three platforms reach universal penetration, everyone in the population is a member of all three (that is, the probability is one).

It is likely that the relationship between members of different platforms is not completely random in this way. For instance, early adopters tend to buy in early to all new technologies; and there is known to be higher PC penetration in digital TV homes, for instance (presumably an income-related effect). So this way of assessing eCRM potential probably somewhat underestimates the reality, at least in the early stages of an evolving digital market. However, by and large, the digital TV, PC/Internet, and mobile telephony markets surveyed have moved out of the early adopter phase.

Moreover, since we are applying the convergence factor in a similar way across all the territories studied, it should not matter too much as far as generating comparative figures is concerned.

Figure 2.7: Convergence Factor in Western Europe 2001

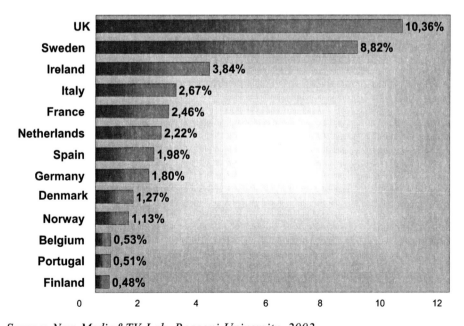

Source: New Media&TV-Lab, Bocconi University, 2002

The Relevance of Interactivity

Another important element in assessing eCRM potential is the extent to which the digital networks facilitate customer tracking. Four levels of interactivity are considered (see Chapter 5 for a deeper analysis of the concept of interactivity):

- local,
- one-way,
- two-way (low),
- two-way (high).

Networks exhibiting a high level of two-way interactivity are obviously those where eCRM potential is greatest. In general, digital TV networks offer a lower level of interactivity than mobile and PC/Internet ones.

The interactivity factor for a territory is calculated according to the following formula:

[(Penetration of digital TV) + (Penetration of mobile telephony x 2) + (Penetration of PC Internet x 2)] / 5

Figure 2.8: Interactivity Factor in Europe 2001

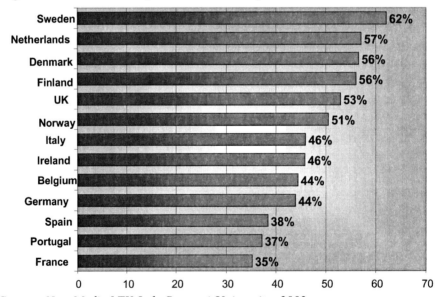

Source: New Media&TV-Lab, Bocconi University, 2002

The Convergence Index

The convergence index is generated as follows:

[Critical Digital Mass index x (1+ convergence factor) x (1+ interactivity factor)]

[(D+M+I)*(1+D*M*I)*(D+2M+2I)/5)]

---

where D = digital TV penetration, M = mobile telephony penetration, and I = PC Internet penetration.

This index represents the "critical digital mass of consumers." It is possible to derive estimates of the number of consumers likely to be present across all three platforms by the simple expedient of taking the population of each territory and multiplying it by the triple platform penetration factor. It is also possible to give an indication of the number of consumers likely to be present across two platforms by doing a double-platform penetration calculation.

Sweden and the UK emerge as the territories with the highest eCRM potential, while Portugal and France are the territories with the lowest.

The countries at the top of the table owe their position largely to high penetrations of mobile telephony and Internet, rather than to digital TV. The UK is a notable exception.

Figure 2.9: Convergence Index in Europe 2001

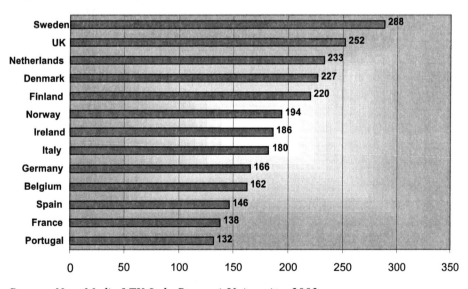

Source: New Media&TV-Lab, Bocconi University, 2002

The conclusions that the model generates are designed to give companies guidance as to how the broad convergence picture will evolve over time in each of the European countries studied, so that they might fine-tune their investment strategies. The goal of this model is not to obtain the accurate size of these cross-platform populations. This model is also a good starting point to address other related questions and allows for further research to profile some of the key players in each territory and channel, in order to assess which types of company are best placed to exploit this newly defined type of convergence.

For companies looking at their digital investment strategy in Europe and seeking to maximise their benefits from consumer convergence, it is key to know which territories exhibit the best potential for development.

Companies knowing this can decide where to initially test or introduce eCRM systems, or how to assess the likely success of existing European projects in an eCRM context.

New markets emerge for the media industry to share in. Conventional media products are in danger of being replaced. Traditional media companies are under pressure to exploit upcoming technologies before newcomers or companies from the IT industry break into their established markets.

CONCLUSIONS

After having dealt with the concept of convergence, this study will focus—in the following chapters—on the analysis of two convergent markets: digital TV and online services, for the purpose of understanding how these two worlds which used to be distinct and far from each other are gradually converging, and which are the new challenges and opportunities all operators involved are faced with.

In the following chapters, the gradual interaction of TV services with potentials offered by the development of online services highlighting interactive television perspectives will be investigated.

From the perspective of online services, interactive television seems to be an incremental innovation that takes the form of broadband online services, whereas from the TV industry perspective, online services were initially seen as a test market for interactive TV.

Figure 2.10 illustrates fundamental innovation paths from the perspective of TV stations and online service providers. Target groups can vary in terms of their homogeneous interest or the hardware platform (device) they use. The concept of media services varies in terms of the hardware platform used or the

Figure 2.10: Innovation Paths Between Television and Online Services

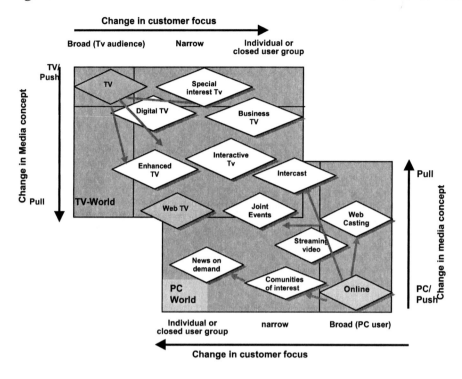

Source: Adapted from Dowling, Lechner & Thielmann, 1998

level of interactivity (variation of push versus pull elements). At this point it is necessary to make a clear distinction between these hybrid forms of products or services, whether they only transfer existing functions from one to another platform or whether they add new functionality through the combination of both media services.

Will online services offered by TV stations be able to boost the TV industry or will TV programming integrated into online services be able to increase radically their growth? A need for convergence is only felt on a limited basis yet; the functions and use of the two media are (at least for the time being) too different.

The portability of contents from one class of devices to another is difficult to achieve due to significant technical differences. Therefore, new devices and transport media must probably be supported with target media specific contents.

Starting from conventional publishing, technical innovations also enable new kinds of distribution (for example, Video-On-Demand) as well as sources of income (for example, Pay-Per-View).

Additionally, interactive online technologies provide completely new types of services based on applications, transactions, or user-driven communication. Establishing these kinds of services presents major challenges for the media industry, because they draw media companies away from their traditional, content-oriented activities.

ENDNOTES

[1] See also OECD. (Ed.). (1992). *Telecommunications and Broadcasting—Convergence or Collision?* Paris, France: OECD.

[2] See Shannon, C. E. & Weaver, W. (1949). *The Mathematical Model of Communication.* Urbana, IL: University of Illinois Press.

[3] Rawolle, J. & Hess, T. (2000). New digital media and devices—An analysis for the media industry. *Journal of Media Management*, 2(2), 89.

[4] Rawolle, J. & Hess, T. (2000). New digital media and devices—An analysis for the media industry. *Journal of Media Management*, 2(2), 89-90.

[5] Blödorn, S., Gerhards, M. & Klingler, W. (2000). Fernsehen im neuen Jahrtausend—ein informationsmedium? *Media Perspektiven,* (4), 173.

[6] Rawolle, J. & Hess, T. (2000). New digital media and devices—An analysis for the media industry. *Journal of Media Management*, 2(2), 89-90.

[7] Pull services typically address only one end-user (the one who triggered the transmission).

[8] Noelle-Neumann, E., Schulz, W. & Wilke, J. (1999). *Publizistik Massenkommunikation.* Frankfurt, Germany: A.M. Fischer.

[9] Customer Relationship Management (CRM) can be described as the process of attracting, retaining, and capitalising on customers. CRM defines the space where the company interacts with the customers. At the heart of CRM lies the objective to deliver a consistently differentiated—and personalised—customer experience, regardless of the interaction channel. Flynn, B. (2000). *Digital TV, Internet & Mobile Convergence—Developments and Projections for Europe.* Digiscope Report, London, UK: Phillips Global Media, p.43.

[10] According to simple probability theory, if we flip a coin, the chance of getting heads is one in two (0.5); if we flip a coin twice, the chance of getting heads twice in a row is one in four (0.5 times 0.5), and so on.

PART II

MULTIMEDIA AND INTERACTIVE DIGITAL TELEVISION

Chapter III

Digital Television

INTRODUCTION

With the introduction of digital technology in the production, distribution, and reception of the TV signals, an actual technological discontinuity occurred which—starting from the first half of the '90s—has been putting pressure on the TV system thus originating an important transformation. The passage to digital TV is a technological phenomenon of the widest range from the standpoint of transmission capability, of efficiency of distribution networks, of image quality, and of flexibility and variety of performances which for the very first time widen the TV set's field of use well beyond the programs' traditional fruition.

But this evolution's value is not only technological. Indeed, it has a profound impact on the entire TV system: from the offer typologies to the consumption manners, from the technological and productive structures to business models. The market outlook, the operators' typologies, the distribution systems are changing. The audience behavior and the TV watcher's status are changing, as well as the nature of the medium and its function.

In the previous section we described the evolution directives featured in the reference scenario. We will now analyse digital TV's technological features with special reference to the advantages and limitations of the different manners of transmitting the digital signal.

Copyright © 2003, Idea Group Inc. Copying or distributing in print or electronic forms without written permission of Idea Group Inc. is prohibited.

WHAT IS DIGITAL TELEVISION?

The term digital television was adopted by the FCC to describe its specification for the next generation of broadcast television transmissions.

Digital television is based on the transmission of a digitised signal which is transformed into a binary numerical sequence, that is a succession of 0 and 1.

To better comprehend digital TV's technical characteristics, we deem it appropriate to describe the digital transmission system by subdividing it into some phases.

Digital Coding

Digital transmission's first technical phase is the passage to digital, i.e., the transformation of the analog signal into digital signal (in a sequence of numbers 0 and 1).

Television operates in much the same way as a movie film. There is a series of frames, each a fraction of a second apart, and each a frozen snapshot of the action seen by the camera. When the frames are played back at the same rate, the human eye is fooled into interpreting the result as a smooth, continuous moving image. Whereas each frame of a movie film really is a photographic snapshot, each frame of a TV signal is several hundred horizontal lines, each representing a scan across the frame. In both a video camera and a TV set, an electron beam or the equivalent literally moves extremely rapidly across the screen, line by line (or sometimes alternate lines), until it reaches the bottom, and then repeats the process from the top, frame by frame.

Suppose there is a single frame (one "screenful") of video that must be transmitted through some communication channel, whether over the air or via cable. If the frame conforms to current broadcast standards, it is an analog image that has been scanned at up to 30 frames per second, with 480 lines per frame. Sending this signal in the usual way over the airwaves or on cable occupies 6 megahertz of bandwidth.

The first step in the compression of such an image is to digitise it. That means taking samples, as described earlier, at specific points—snapshots of the state of the analog image at periodic intervals. Each sample generates what is called a "pixel"(picture elements)—a small square or rectangle representing the information from the sample at a particular point[1] (see Figure 3.1). The more sampling there is per line, the smaller the pixels, and the finer the "grain" of the resulting image when received.

Figure 3.1: Digital Video Compression

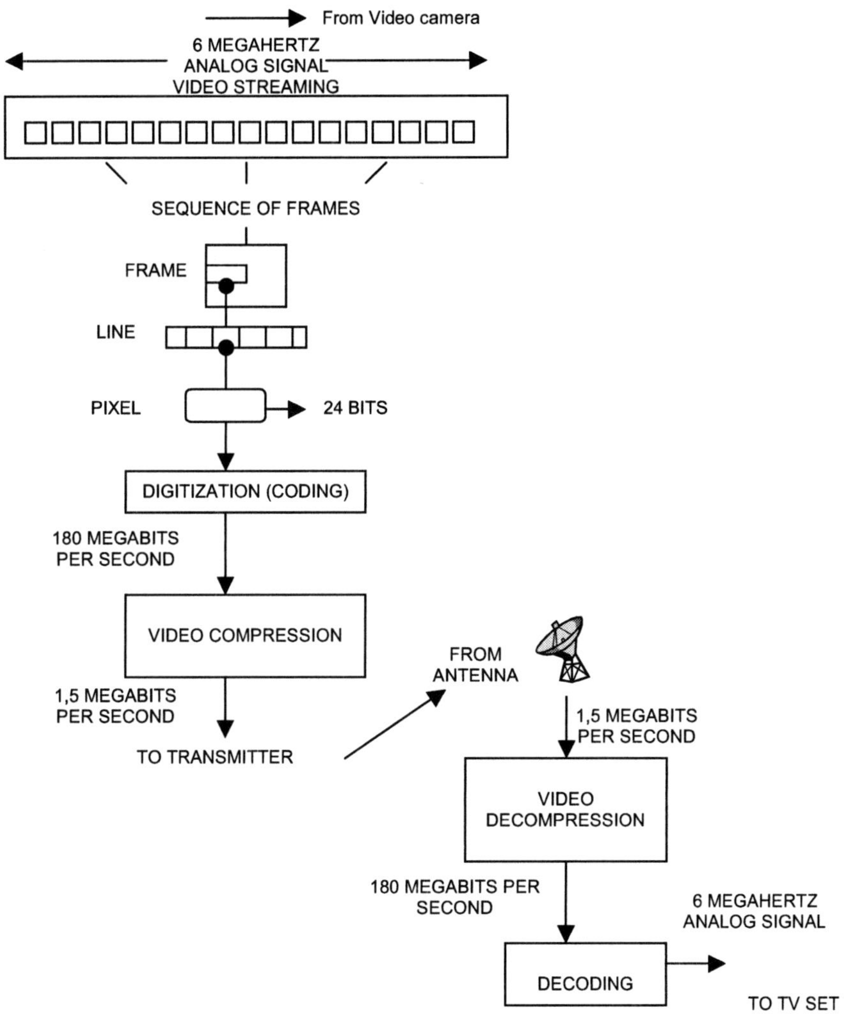

Source: Adapted from C-Cube Mycrosytems, 1996

Digital Compression

The second passage is the compression of the digitised signal adapting it to the transmissibility and filing economy requirements made necessary in view of the now available technologies. This technique's common principle is the application of computer algorithms to the digitalized signal for the purpose of reducing space and time redundancy. During such phase the signal is compressed with a resulting saving of "space" and with a more efficient use of the bandwidth: on a typical 33 Mhz transponder—where now only one analog channel can be transmitted—a digital load of 55 MB/s, equal to roughly 10 digital channels, can be housed.

That means the information in five or six TV channels is squeezed into 6 megahertz and the signal can be transmitted on a narrower bandwidth with the same or better reception.[2] There is no one ratio of normal compressed channels because the degree of compression that is possible depends on the amount of redundancy in the uncompressed signal. Channels with lots of action cannot be compressed as much as channels with less action. Material broadcast "live" in real time cannot be compressed as much as material that can be run several times through a processor.

The transformation of the normal analog signal in a sequence of numbers (0 and 1) and the related compression guarantee a savings not only in terms of space occupied on the transponder, but also in terms of information relating to the single programs.

During compression of a TV event, a special device called an *encoder* compares each frame with the next one: in the passage from one to the next, only the differing parts are transmitted, leaving the unchanged ones "unprocessed" or "untouched," thus allowing huge bandwidth savings. The transformation into numbers is advantageous in more ways than one, certainly not only quantity-wise. The digital flows relating to each channel contain audio and video information, as well as appropriate mistake correction codes which permit elimination of any "noise" due to transmission. In other words, the quality of program reception in digital is always perfect, never static-affected.

The audiovisual signal's encoding algorithms establish the manner of application of one or more signal compression techniques and the rules to transmit it. The encoding of audiovisual sequences shall contain—besides information merely descriptive of the sequence content—the control information required for the reception device to correctly interpret the signal.

The most popular current standards for video compression are ones established by a voluntary worldwide industry association, the Motion Picture Experts Group (MPEG), and adopted by the International Standards Organization (ISO). The first of these, MPEG-1, introduced in 1992, is tailored for use in connection with compact discs that contain video images. The more recent MPEG-2 standard has been adopted by the direct broadcast satellite industry and by the FCC for digital television. MPEG-2 is designed to accommodate high-definition television with a wide screen and digital sound. MPEG-2 uses "interframe" coding, transmitting only the pixels that have changed from one frame to the next. In addition, some redundant information within frames is removed, along with certain aspects of each image that have minimal visibility to the human eye, such as vertical and diagonal movements.

MPEG-4 provides for video transmission over narrowband channels, that is telephone wires.

By now the compressed digital signal can be transmitted (Phase 3). Several manners of transmission are possible—like for traditional television, i.e., via airwave, cable, DTH (*Direct To Home*) and DBS (*Direct Broadcast Satellite*) satellite, ASL (*Asymmetric Subscriber Loop*), and MMDS (*Multichannel Microwave Distribution System*)—which present differences at the level of means of signal production and reception and—only marginally—at the level of transmission system.

ADVANTAGES OFFERED BY DIGITAL TELEVISION

Digital-TV-originated advantages are in any event plentiful thanks to the possibility of compressing each signal, which is conducive to a much wider choice of TV and radio channels.

The reduction in the use of the electromagnetic spectrum, due to digital signal compression, produces a subsequent increase in the number of channels which can be transmitted and an increase in choice options. With the same quantity of frequencies needed for an analog TV channel, we can now have four to six digital channels (a number very likely to increase in the future). For example, in the electromagnetic space occupied by one airwave analog channel, today we can transmit 18-24 Mbits per second, while transmission of a standard definition digital channel requires a capability of about 4 Mbits per second.

As we move from analog TV to digital, there is increased channel capacity that can be cleverly used in a number of ways.

Change in the nature of the signal has had substantial effects such as:

- better picture and sound quality (better reception and higher quality in signal diffusion) since numerization engenders a higher degree of immunity to distortions and disturbances introduced in the channel via air—thus guaranteeing a steady diffusion quality value;
- possibility of exploiting bigger TV screens (from the 16:9 screens to big flat ones);

- more flexibility for the broadcaster in using transmission resources—for example, in a certain coverage area, we can decrease the number of broadcast channels in exchange for higher image quality which can be broadcast even in High Definition TV (HDTV);
- integration of Web technologies with digital TV such as transmission of Web information, development of interactive, and Near-Video-On-Demand (NVOD) services[3];
- increased programming options (e.g., multiple camera shots used in sports telecast);
- security issues—digital encryption for scrambling programming such as Pay TV services in order to make service inaccessible to "pirate" non-paying viewers;
- easy integration among broadcasting networks and broadband telecommunication networks (e.g., B-ISDN);
- spectrum auction and local government use.

Furthermore, digital technology makes the TV medium more flexible. The user has ample choice of programs, to the point that he can create his own personalized program set.

Also media integration increases allowing combination of the TV services Web technologies such as Internet surfing, as well as the carrying out, in the comfort of one's own home, home banking operations, TV shopping, or playing games.

As things stand, we can benefit from some typologies of digital TV channels:

- digital channels which broadcast without any variation the analog networks' programs;
- *multiplexing*—the program is broadcast live on different channels but in different manners (on different channels the broadcast may be the same soccer match with different shots and directions);
- *theme channels* specialise on theme sets of programming such as sport, movies, information, or others;
- *Pay-Per-View*—programming is not paid entirely by the user, but only based on actual viewing. There's no fee to be paid by the user, only the fee related to the specific program (e.g., the movie or the soccer match) chosen;
- *Near-Video-On-Demand (NVOD)*—the same broadcast is repeated at very close time intervals (e.g., a movie is broadcast each half hour of the

evening on different channels). This requires use by the broadcaster of more channels for the same program.

Further services offered by the digital TV channels are added information services destined to specific target audiences, as well as simultaneous broadcasting of the movie either in the original language or in the dubbed version.

Furthermore, digital technology allows development of interactive programming. Interactive television services include both diffusive numerical services (Pay-Per-View, Near-Video-On-Demand), and asymmetrical interactive video services[4] (TV banking, TV shopping, interactive games, etc.). Technically, interactivity implies the presence of a return channel in the communication system, going from the user to the source of information.

In light of these radical technological transformations and of the new market's needs, the European Community created in 1993 the DVB (Digital Video Broadcasting) project aimed at defining an organic strategy for introduction of new numerical services of TV radio diffusion.

TRANSMISSION SYSTEMS

The digital television signal can be transmitted via several transmission media such as satellites, cable networks, terrestrial broadcasting channels, and other media: MMDS (Multipoint Microwave Distribution System) and ADSL (Asymmetric Digital Subscriber Line).

It bears pointing out that bit transmission can take place via two main modalities:

1) via wire, which in turn can be either the traditional copper coaxial wire or the optic fiber innovative one;
2) wireless, which means using airwaves or the satellite.

The different transmission means (wire or wireless) are at the disposal of different operators, whatever the sector contest to which they belong.

All the DVB radio-TV systems have some common phases such as:

- Phase 1: image and sound encoding[5] and multiplaction or regrouping/combining[6] by packages, a common system to supply the user with information on service configuration (SI – Service Information);

- Phase 2: external adapter for error protection;
- Phase 3: channel adapter (internal encoding and modulation[7]).

These common elements guarantee a high level of compatibility among the commercial type IRDs—integrated receivers/decoders—for the various transmission media.

Each transmission support has specific characters which shall be analysed in order to point out their respective advantages and limitations.

Digital Satellite Transmission Systems

Nowadays, DTH (Direct To Home) satellites are being used which are largely employed for diffusion of analog TV signals. For such applications a satellite is normally subdivided into a certain number of "transponders" which may be considered, for all intents and purposes, as amplifiers. A transponder, except in some cases, has a bandwidth calculated to be assigned to a single analog TV channel.

A typical configuration for the satellite distribution of four numerical TV programs according to DVB standards is shown in Figure 3.2. The four source

Figure 3.2: Typical Satellite TV Transmission

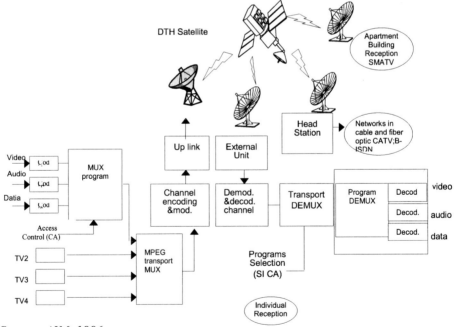

Source: AIM, 1996

signals encoded in numerical (MPEG 2)[8] and probably crypted[9] for access control (AC) are sent to the transport MPEG multiplactor, which then builds the serial numerical flow.[10]

Such numerical flow, upon exiting the channel codifier, is sent to the earth station which then takes care of transmission to the satellite. Reception of the irradiated satellite signal can occur through individual receiving systems, SMATV systems for apartment buildings reception, and CATV (cable networks).

In the case of satellite transmission, the transmitted signal is picked up by the satellite dish and conveyed on to the LNB (Low Noise Block), an electronic device on the dish's focus which amplifies and converts certain frequencies into others located in a lower band of the electromagnetic spectrum; thus, the received signal can be fed into the coaxial cable which carries it inside the home.

Finally the signal arrives at the IRD (Integrated Receiver Decoder), a last generation set top box, equipped like a personal computer, which decodes the signal. These new receivers turn the digital signal into an analog one, to adapt it to the TV set. Furthermore, they decode the crypted signal of pay programs and, via insertion of the smart card in its slot, activate access to encoded channels.

For fruition of the new digital Pay TV, the user shall have a satellite reception system: a small diameter dish, a universal converter, and a digital receiver based on the MPEG2-DVB European standard.

Besides being connectable via modem, IRDs also have an outlet to connect to a Hi-Fi set which enables fruition of a sound characterised by extraordinary compact disc quality.

It is technically possible to utilize as a return channel the phone line linked to a Service Center which sends data to the satellite, but in practical terms there is no chance of serving a single user using a portion of frequencies which could bring light to a great expanse of a whole continent. The return channel, made possible through the phone line or a Service Center, is used to run PPV (Pay-Per-View) services: in this case the Service Center sends the control signals to the satellite which enables users to receive chosen programs (Conditional Access).

Digital Cable Transmission

Unlike satellite and terrestrial-based systems, a cable system appears to contain an inbuilt return-path, since a wire is laid in the ground that physically connects the subscriber to the service provider. For this reason, cable often boasts that it is an inherently two-way medium—in a way that satellite and terrestrial are not.

In fact, cable networks have, historically, principally been built as simple re-transmitters of broadcast programming, in which a multi-channel broadcast analog signal is sent out, split, amplified, split again, and amplified again until it reaches the subscriber.

This type of network is inherently one way, and has in practice proved expensive to upgrade to two-way capability. Even today, in the world's largest and most mature cable territory, the United States, it is estimated that only 68% of all cable homes are passed by activated two-way plant—after a program that has cost many billions of dollars.

However, once a cable network has been upgraded, it is clear that the cable return-path is capable of coping with much higher speeds in both directions than either of its rivals.

Generally, one or more 8MHz chunks of bandwidth are typically set aside for two-way applications, which—in a digital environment—are addressed through the use of a cable modem in the set-top box.[11] A realistic bit rate is probably between 128 Kbit/s-500Kbit/s upstream under typical subscriber loads at peak time. Meanwhile, maximum throughput downstream is between 38 Mbit/s and 50 Mbit/s per 8 MHz channel depending on the modulation scheme used.

Cable platforms offer return path speeds up to 10 times as fast as either terrestrial or satellite digital set-top boxes, relying on analog dial-up modems, whichever cable modem standard is chosen. This means that they support high-level, two-way interactivity much more easily. At these rates, for instance, good quality videoconferencing would be eminently possible between two cable subscribers on the same network, as would such applications as fast multiplayer gaming between different cable homes.

Digital Terrestrial Television (DTT)

Terrestrial TV networks rely on over-the-air transmissions from a transmitter site to a rooftop or indoor aerial. Transmitters radiate at high power in all directions, and are therefore able to reach most, if not all, of the aerials in their locality with a useable signal. Sending a signal back in the opposite direction from the aerial to the transmitter is a much more difficult proposition: it would have much lower power, and the signal would therefore have to be very precisely aligned to a receiver on a transmitter mast.[12]

DTT operators wishing to provide their viewers with a return-path do so by inviting them to plug their digital set-top box into their domestic phone line.

An analog modem built into the DTT box is then able to transmit signals back to the service provider in the same way as a standard PC dial-up connection to the Internet does.

The interactivity afforded in a DTT environment is relatively crude. One reason is related to early DTT set-top boxes which, in order to be affordable, have limited storage and processing power (unlike PCs). This constraint is all the more acute because of the higher cost of DTT boxes, which require a more expensive demodulation chip than satellite or cable ones.

ADSL (Asymmetric Digital Subscriber Line) Transmission

ADSL is a transmission technology that turns an ordinary telephone line (often referred to as twisted copper pair) from a low-capacity environment (64 Kbit/s in either direction) into a medium-capacity one. The technology relies on installing a pair of modems at either end of the local loop (one in the subscriber home, the other at the local exchange or central office) which modulate and demodulate the transmitted signal in such a way that its carriage through the copper wire is optimised. The voice signal is carried in a different part of the copper spectrum, and is unaffected by the addition of the ADSL modems.

Many local telephone companies are experimenting with these technologies, chiefly "asymmetric" DSL (ADSL). Asymmetric indicates more data flows in one direction (downstream, to the user) than the other.

ADSL performance levels tend to be affected by the quality of the network infrastructure in the territory concerned: if the quality of the network is poor, speeds will typically be lower than these—or ADSL service may be impossible. Within these constraints, service offerings are also affected by the extent to which there is a requirement for universal service. In the UK for instance, where telecommunications regulator Oftel requires incumbent telephony operator BT eventually to make ADSL service available to over 85% of households, BT believes it can only guarantee around 2Mbit/s downstream to this proportion of homes (less than a thousandth of the capacity of a digital cable network). Homes close to exchanges would presumably be able to access higher rates if offered, but BT must cater to the lowest denominator as a common carrier, not the highest.

Table 3.1: DSL Variants, Speeds, and Distances

DSL Variant	Meaning	Speed range	Range (m)
ADSL Lite (G Lite)	Asymmetric Digital Subscriber Line	Up to 1.5Mbit/s downstream (VHS quality video) and up to 384 Kbit/s upstream	7,100-8,100
ADSL	Asymmetric Digital Subscriber Line	Between 1.5 – 8Mbit/s (broadcast quality video) downstream and up to 640 Kbit/s upstream	3,900-5,800
HDSL	High data rate Digital Subscriber Line	1.544 Mbit/s full duplex or 2.084 Mbit/s full duplex (VHS to broadcast quality video)	3,900-4,800
SDSL	Single line Digital Subscriber Line	1.544 Mbit/s full duplex or 2.084 Mbit/s full duplex (VHS to broadcast quality video)	3,200
VDSL	Very high data rate Digital Subscriber Line	13-55 Mbit/s downstream and 1.5-2.3 Mbit/s upstream (several channels of broadcast quality video)	300-1,500

Source: ADSL Forum

For these and other reasons, service offerings tend to vary greatly between countries. In research carried out for the UK regulator Oftel,[13] analysis found that for services targeted at residential users, maximum downstream bandwidth offered ranged from:

- 256 Kbit/s to 2Mbit/s in the UK (broadcast quality video at best);
- 380 Kbit/s to 1.5 Mbit/s in the United States (VHS quality video at best);
- 500 Kbit/s to 512 Kbit/s in France (1/4 screen broadcast quality video);
- 768 Kbit/s in Germany (1/2 screen VHS quality video).

If ADSL were to be used in digital TV broadcast mode, the available capacity would offer at best a couple of high-quality video channels at any one time—worse than the performance of a terrestrial analog system. ADSL is therefore used in on-demand mode for digital TV.

The viewer makes a request via the remote control that passes up the network to a central server where a library of digitised TV programs or films is stored, and the server immediately streams the video directly to the viewer's modem and TV set.

It is possible to use this type of model to provide the experience of a multichannel broadcast TV network by taking live feeds of broadcast channels and storing them on the fly on the remote server in such a way that they can be re-presented to the viewer virtually simultaneously with the original broadcast. What the viewer is effectively doing in this situation is to get the remote server to switch channels one at a time onto his or her personal "pipe."

There is still an on-demand model in which there is a one-to one link between the viewer and the service provider, but as far as the viewer is concerned, the system provides the illusion of simply switching between different channels locally.

The efficiency of this type of system can be improved by using techniques such as IP Multicast, in which a single copy of a program stored on a server can be duplicated further down the network at router level to satisfy multiple simultaneous requests.

In most European countries, ADSL is regarded as technology for delivering fast Internet access to PCs rather than on-demand video to TV sets. A notable exception is the UK, where most of the significant early players in the ADSL market plan to focus on offering Pay TV. Even incumbent telephony operator BT is providing a "video-rich" Internet portal. This partly explains why speeds offered in France and Germany are lower than in the UK: operators simply do not see the need to offer speeds this high if they are not going to provide broadcast-quality video.

Since ADSL is in its early stages in Europe, it is difficult to gauge practical performance. There is increasing evidence from the United States that once a telephone exchange is carrying a penetration level of around 40% or above of ADSL modems, serious problems begin to be encountered with interference and "cross-talk." However, below these penetration levels one would in general expect ADSL to be capable of operating much closer to its maximum theoretical limit than cable modem technology, since the capacity is not shared between subscribers but dedicated to each telephone line.[14]

MMDS (Multichannel Multipoint Distribution System)

Technologies for signal transmission on microwave carriers are an alternative to cable in the laying of wide-band networks and—primarily in areas with a low/average population density—they are a rapid way to carry service wherever legislative barriers are in force for the digging needed to lay a cable.

MMDS (Microwave Multipoint Distribution System) is a system working on the electromagnetic spectrum's lowest frequencies (2-10 Ghz).

This *"wireless cable"* system employs very short waves and one or more antennas from which the signal originates, which arrives straight to the users and are placed at such a height as to guarantee going over the obstacles which may be present in the area to be covered. The MMDS microwave system generally has a covering radius of about 20-25 kilometers if omnidirectional, and about

50 kilometers if semi-circular, thus being particularly fit for the multimedia service offered.

We should point out that the functional scheme of the microwave systems' reception device is conceived in such a way as to utilize the set top box employed in satellite or cable transmission systems.

The first element of the reception system is therefore an antenna positioned on the building roof in the transmitter's line of sight. Once picked up, the signal is converted to a lower frequency by a specific *front end* and then sent to a standard set top box. Just like for other digital transmission systems, complying with the DVB standard, the set top box carries out signal demultiplaction, decompression, and decoding functions, and the signal is finally sent to the TV set.

One of the first projects using MMDS (*wireless cable*) transmission technology in Europe is the one carried out by *Irish Multichannel* with *Off Air Electronics* in Ireland for broadcasting—by April 2000, 60 TV channels were being transmitted to 120,000 subscribers at a total project cost equal to about $6.5 million.

In the U.S. a few of the large regional Bell telephone companies have toyed with MMDS as a means to compete with cable. PacBell has converted several MMDS systems in California to digital transmission at a cost of more than $150 million, greatly increasing channel capacity, and it appears ready to compete vigorously with cable in those areas.

As it has become more apparent that cable companies are not immediate competitive threats in the telephone business, telephone company interest in video competition, via MMDS or otherwise, has waned. Apparently telephone companies had little interest in offering video service as a separate line of business. Instead, they saw video investments chiefly as counters to anticipated cable entry into the telephone business.

One reason for the lack of success of MMDS relative to cable television is limited channel capacity.[15] Although it has increased, the channel capacity lags behind that of cable and direct broadcast satellites. Moreover, the frequency band available to MMDS operators is more limited than the bands available to other wireless service providers. Another handicap has been the need to install special antennas on subscriber rooftops, along with electronic equipment to decode the signal. The cost of the special equipment and installation offsets most or all of the advantage of not having to run wires down the streets. Finally the microwave bands are subject to degradation due to rain and other factors, causing reception problems.

LMDS (Local Multichannel Distribution System)

LMDS operates at 28 and 31 gigahertz, a frequency band (the "Ka-band") shared with GEOs (Geosynchronous Earth Orbits)[16] and LEOs (Low Earth Orbits).[17]

LMDS is a cellular service of necessity: at 28 gigahertz, signals are quickly attenuated by terrain and atmospheric moisture. But a cellular structure is also an advantage because it multiplies capacity through spectrum reuse. LMDS also permits "bandwidth on demand." To provide service to a new customer in the coverage area, only subscriber end equipment need be installed. If the system is interactive in the Video-On-Demand sense, bandwidth can be allocated to individual users in a cell as needed, yielding savings through time sharing of frequencies; there are no dedicated user channels, as with telephone-based systems. In comparison with local telephone loops and cable drop lines, this provides the opportunity for substantial economies.

Finally, the technology may permit an interactive service in the sense of video telephony or broadband Internet links; this requires more sophisticated and expensive transmitting equipment for each subscriber.

An experimental LMDS license, the first, was awarded to CellularVision in 1991. CellularVision built and operated an analog LMDS system in portions of New York City, offering one-way video service.

RECEPTION SYSTEMS

After having dealt with the diffusion of the TV signal via different media, it seems appropriate to explain how the signal is received by the user.

The digital signal reception by the user (fourth phase) is made possible through a digital adapter (henceforth called *set top box* or *decoder*). This device constitutes the primary interface between subscriber and the numerical services' distribution system. Of a size similar to a VCR, the device is positioned close to the TV set and, once connected to the network and the TV set, it enables the user to both receive and select, via remote control, the services being sent (although at present the set top box is an "add-on" product, in the future it will be integrated inside the TV set itself).

The set top box has the function of decoding digital signals to enable them to be interpreted by the old analog TV set, and has processing and memory capability which allows it to process and store information. However, there are different variations of set top boxes to be taken into consideration, such as the one for satellite, for cable, for ADSL, and for cellular television.

The device has a key role for diffusion of the new numerical video services: it will have to carry out a relevant number of new functions to run and manage numerical communication between the service center and the viewer, and between the viewer and the network to select and process the audio-video signal relating to the chosen service. Furthermore, it will require the capability of managing the signals sent to the service center for access control.

The commercial success of numerical multimedia services destined to big audiences requires the short-term lending of a set top box to the viewers to link nowadays to home TV sets, at costs already included in the services' launching phase. The need for the signal to be filtered by such devices has made the set top box a strategic element. That is why, in spite of the DVB project's efforts aimed at generating a market uniformity based on a common standard, digital TV operators tend to build closed proprietary systems, each with the objective of developing a market through the distribution of the devices which are a must to make the system work and which make up the so-called "digital platform." A good example within this framework is the comparison in Europe between the SECA (*Société d'Explotation du Controle d'Access*) system, created by Canal Plus and Irdeto, manufactured by Nokia, Grundig, Italtel, National Panasonic, and Pace, used by Stream subscribers.

The encoding systems which are used are divided in two main categories. The first one includes systems of the "proprietary" kind in which the *service provider* tries to defend the whole value chain of the entire system. By doing so, the moment one buys the set top box, one becomes automatically tied to the supplier or broadcaster of certain types of programs. Then there is the set top box one can buy in any store, open to different possible service providers, characterized by a simple plate or card which can be inserted via the so-called connector or PCMCIA interface (the same interface you use in a computer when you introduce a modulator), which characterizes the type of encoding used by the specific service provider. The smart card, looking exactly like a credit card, characterizes the service profile required by that particular service provider.

General Functions

The numerical set top box is a system capable of carrying out a series of well-defined functions:

- interface to the transmission medium or network interface: this sub-system is a receiver specialized for the particular transmission medium being used (satellite, cable, coupler, etc.) and its task is reconstructing in the correct way the sequence of bit received (tuning, demodulation, correction of errors due to transmission);
- identifying and selecting the type of data received (demultiplexing);
- carrying out control of authorizations for service access and decoding of "pay"-type programs (*"Conditional Access"* and *"Descrambling"*);
- image and audio decompression as well as specific functions on data (*decoding*);
- interactivity-related problem solving;
- running the user interface in order to allow easy surfing in the plurality of services offered (EPG—*Electronic Programming Guide*).

It is therefore interesting to examine at a deeper level herein three specific functions of the set top box:

- *"Conditional Access"* or CA
- *"Return Path"* or *"Return Channel"* (RC)
- *"Electronic Programming Guide"* or EPG

Conditional Access

A system of *"Conditional Access"* (CA) is used inside the set top box to control access to diffusion services whatever the diffusion channel (satellite, cable, etc.). It includes the combination of two functions—the first one is *scrambling* in order to make audio, video, and data information unintelligible, and the second one is *encryption* which prevents unauthorized receptions.

The primary purpose of a conditional access system is to decide which set top boxes are set up to receive predetermined program package (Pay TV) or a specific program (PPV—Pay-Per-View) for the chosen duration. There may be several reasons for this control, such as authorization after payment has been made, limiting access to specific geographical areas, parental control, etc. Besides these functional requirements a primary purpose of CA systems is to abate acts of "piracy" which, by the selling of illegal receivers, are capable of descrambling the signal, thus nullifying conditions of payment for Pay TV and Pay-Per-View to the service provider that has bought the rights to broadcast the programs.

Return Path

The *Return Path* or Return Channel (RC)—also called "interactive channel"—is the means by which the user communicates with the Service Provider and interacts with the Service Center.

One can fully appreciate the importance of such a tool if the focus of attention is shifted from mere diffusion services such as traditional TV and Pay TV to more elaborate services such as "Pay-Per-View," "Video-On-Demand," etc., where there must be interaction with the user and therefore an "*upstream*" communication from the set top box to the Service Provider's Service Center.

One need only consider VOD (Video-On-Demand), which is by now the world's most widespread and popular interactive service through which the viewer can build his own personalised program schedule by "surfing" in the Service Provider's system. For such surfing, but first and foremost for ordering a program, the viewer must be able to carry out a series of commands (generally via remote control), which in turn must be transmitted via the return channel in order to be able to meet the user's requests and keep consumer data updated within the Service Provider's system—with particular reference to billing.

In more general terms the return channel enables viewers to:

- request the "Pay-Per-View" programs and report consumption, making billing automated;
- endorse interactive TV forms such as votings, games, Video-On-Demand, TV shopping, TV banking, etc.;
- monitor the signal reception quality;
- confirm that the "*Conditional Access*" messages sent by the Service Provider have actually been received in order to avoid time-consuming and useless repetitions on the transmission medium being used.

However, there are some disadvantages working against implementation of the return channel, first among which is the terminal's higher cost.

All in all, the creation of a point-to-point link between Service Provider and each set top box is highly recommended since it increases both the security and the performance of the whole system, not to mention allowing for the offer of more evolved systems.

Electronic Programming Guide

The *Electronic Programming Guide* (EPG) is an essential, navigational device allowing the viewer to search for a particular program by theme or by category and order it to be displayed on demand. They are currently the highest traffic areas on all platforms and are therefore of great interest to advertisers.

EPG is a tool offering the viewer a simple way to:

- find titles and starting times of events in accordance with certain user-selected criteria (i.e., time, type of program or service, topic, promotional ads);
- automatically access the selected program;
- program terminal switch on for recording;
- send Pay-Per-View request;
- access further info regarding events, trailers, etc.

The amount of EPG info to be broadcast varies according to the number of services being run, the convenience of use of the guide, and the level of detail required, and it can even reach remarkably high levels. This is why planning an EPG must take the following factors into consideration:

- the bit rate increase required by these auxiliary services to increase the required detail level, which cuts down the one available to programs;
- the user's access time;
- the amount and hence the cost of this service's hardware and software.

THE PROBLEM OF STANDARDS

Technology has acted as a shuttle-cock in the digital revolution by introducing processing and transmission systems of the audio-visual signals defined on the basis of a coordinated work on the part of all sectors' operators (DVB—*Digital Video Broadcasting* working group promoted by the European Commission). This approach has led to standardization[18] of a series of common technical norms for transmission in Europe of satellite digital TV signals (DVB-S), SMATV (apartment building satellite), cable (DVB-C), via airwaves (DVB-T), and MMDS (microwave transmission system) based on the DVB-MPEG2 standard.

Up until now these norms have concerned information encoding, signal compression and modulation, error correction, channels' encoding, all the way to the common algorithm for signals' factorising. They have, however, allowed the operators free choice as to definition of technological solutions for viewers' access control to encrypted services (CA systems). This has turned out to be the battlefield, as it were, of operators during the first phase of the introduction of digital TV—characterized, as we have seen, by a strong component of conditional access TV services—with arrangements polarized around a limited number of systems[19]:

- the Seca/Mediaguard standard developed by the Canal Plus French group and the Bertelsmann German group, used in France, Spain, Italy, and Scandinavia by the "bouquets" leading to the Canal Plus groups (in Great Britain it will be used by airwave broadcasters);
- the d-box developed by Irdeto (Nethold) and Beta Technik of the Kirch group, used in Germany by the Kirch group "bouquets" (DF1) and in part also in Italy where, with the arrival of Canal Plus French shareholders in the Tele+ body of shareholders, the Irdeto system has also been flanked by Seca/Mediaguard;
- the Viaccess standard, proposed by France Telecom and used in France by AB SAT and TPS operators.

On the regulatory level, an effort has been made towards guaranteeing diffusion of open and non-proprietary technologies, protecting consumers' investments, and at the same time ensuring free access by all operators to technological platforms[20] operating in the market (use of common interface and the *multicrypt* system instead of the *simulcrypt* system).[21] These positions have ended up conditioning not only the alliance policies among operators but also the operators' commercial strategies vis-à-vis the option of selling or renting the set top box.

Further relevant effects have occurred in the position of home electronics' manufacturers as it pertains to the choice of technologies to be adopted and the starting up of production levels high enough to allow for scale economies—thus a significant drop of their costs for the target user—as well as in regards to introduction timing of the next phase of digital services and the development of interactive applications, allowing even the running of non-television multimedia services.

However, the entire sector is well on its way to overcoming the technological barrier with standard consolidation, allowing introduction of an universal system for digital TV—known as *multimedia home platform (MHP)*. This system will make different access control technologies perfectly compatible and inter-operative, thus overcoming the diatribe between open and proprietary systems. MHP requires definition of some key specifications, including the (common) interface with the operating system of the machine on which interactive applications are based (API, *Application Program Interface*) and in particular the EPG (*Electronic Programming Guide*). Specifications for which definition is under way, based on cross-platform (Java, Mheg) languages, will increase systems' versatility, also allowing development of new interactive applications.

ENDNOTES

[1] See also Barbero, M., Cucchi, S. & Stroppiana, M. (1991). A bit rate reduction system for High Definition Television (HDTV) transmission. *IEEE Transactions on Circuits and Systems for Video Technology,* 1(1), 4.

[2] See Owen, B.M. (1999). *The Internet Challenge to Television.* Cambridge, MA: Harvard University Press, p. 161.

[3] See Chapter 5 for a complete description of new interactive services offered by digital television.

[4] See Chapter 5.

[5] Encoding: defines the manners or modalities in which numerical systems are processed (e.g., compression, protection from errors, etc.) in order to be transmitted to the different transmission carriers.

[6] Multiplaction or regrouping: technique to combine more signals in a single medium.

[7] Modulation: process by which the feature of a carrying wave (e.g., width, frequency, phase) is varied according to the same law governing the signal good for transmission.

[8] MPEG 2 *Motion Picture Expert Group*—audio and video signal compression and encoding standard.

[9] Crypting: technique for data protection based on transencoding of said data in accordance with rules established by the crypting algorithm; thus the data cannot be interpreted but by one receiving party which knows that algorithm.

[10] Multiplactor: device carrying out the multiplaction (regrouping or combining) of different signals.

[11] European operators use one of two cable modem standards. The first is the U.S. cable industry's DOCSIS (Data Over Cable Service Interface Specification) standard—developed initially for stand-alone devices attached to a PC, and therefore with an emphasis on the transmission of IP data only. The second is the European

DVB/DAVIC equivalent, developed primarily for set top boxes, and with a much wider range of services in mind. These have somewhat different characteristics.

[12] This is not an insuperable difficulty, apparently. Irish broadcaster RTE has planned to introduce just such a "wireless" return path with the launch of DTT.

[13] Analysis final report for Oftel. (2000). *International Benchmarking of DSL Services.* London: Oftel.

[14] There is likely to be a difference here between the user experience of digital TV and that relating to fast Internet access, since video content is more likely to be stored on local servers for subsequent play out. See also Flynn, B. (2000) *Digital TV, Internet & Mobile Convergence—Developments and Projections for Europe.* London: Phillips Global Media, pp. 211-213.

[15] Owen, B.M. (1999). *The Internet Challenge to Television.* Cambridge, MA: Harvard University Press, pp. 144-145.

[16] GEO: geosynchronous communication satellite in the Clarke orbit at an altitude of 22,300 miles. It remains in a fixed position relative to the Earth.

[17] LEO: Low Earth Orbit satellite systems. Communication satellites in orbit at altitudes of a few hundred miles. Each system has dozens or even hundreds of satellites, each of which is in constant motion relative to any point on the Earth. *See GEO.*

[18] Technically speaking, a standard is a set of regulations to which manufacturers of single parts must comply in order to guarantee integration of such parts within the system to which they belong.

[19] Ajello, B. (1998). *Le Principali Tendenze della TV Digitale in Europa e in Italia.* In L'Industria della Comunicazione in Italia, Turin, Italy, Guerini e Associati.

[20] Digital platform: a set of technologies and infrastructures by which a broadcaster distributes TV and multimedia services to the users. The signal is processed by a *service provider*, transmitted via a transport network called *network delivery system,* and received by the user.

[21] In the *multicrypt* system the signals' decoding and *descrambling* occurs in a module outside the decoder—to which it is linked via a common interface—in which the conditional access system is located. The decoder can then qualify to screening of different operators' programs by changing the outside conditional access module and the smart card as needed. The *simulcrypt* system used in the presence of conditional access system inside the decoder allows transmission, within the same data flow, of different control messages for authorization of encrypted signals. This way the same program can be received by different decoder families, with different conditional access systems.

Chapter IV

The Economic Implications of Digital Technologies

ADVANTAGES AND LIMITS IN DIGITAL SIGNAL TRANSMISSION METHODS

Having described the technical characteristics of the different digital signal transmission methods, it seems appropriate to investigate the specific characteristics for each medium, underlining both the advantages and limits of each and their notable economic implications. Transition from analog to digital has not only technological effects, but also effects on an economical and cultural level for the television world.

Digital Terrestrial Television

For years television broadcasting has had the advantage of guaranteed coverage across Europe with access in almost all homes (over 95%) through the installation of simple and inexpensive antennae. The superior potential of terrestrial broadcasting in Europe, when compared to cable or satellite, lies in its access to a greater percentage of homes. For this reason in the medium-long period (15-20 years) in Europe, terrestrial networks are considered to correspond best to the needs of a universal or commercial (advertising) service as they guarantee widespread and economic broadcasting of television channels.

Terrestrial broadcasting networks also have other advantages[1] such as:

- *Portability:* With terrestrial broadcasting the television set can receive programs in any location inside or outside the home by means of a mobile antenna (it is not necessary therefore to have access to a connection point at every location of possible television consumption).
- *Regionality:* Geographically it would not be possible to cover the entire region with a cable network (excessive cost) while satellite coverage is geographically extensive and cannot be limited to a regional scale.
- *Technical Vulnerability* of terrestrial transmission is clearly inferior to that of satellite transmission. A terrestrial network is made up of a variable number transmitters and relays distributed over the serviced area. In the event of technical shortfalls, electricity supply failure, or other breakdown, the interruption is limited to the area covered by the transmitter or relay involved and has no effect on the rest of the network. Instead, with satellite transmission any type of failure (in the absence of a backup) implies a complete interruption of the service in the entire area covered.

The principal obstacle to the short-term introduction of digital television in terrestrial broadcast is the shortage of channels available on the present European scene, characterised by a high use of spectral resources by analog television systems such as PAL and Secam.

The introduction of digital television services is foreseen in two phases—the first is short term and foresees a co-existence of both analog and digital; in the second, more long term (about 15-20 years), analog signals will be completely replaced by digital.

While in the first phase it will be necessary to protect the analog signals from interferences to which they are intrinsically very sensitive, in the second phase it will be possible to exploit the advantages of digital technology to obtain the maximum number of channels.

It is worth noting that already in the early phases it will be possible to use, on a local level, the channels not used by analog systems to obtain coverage of limited but densely populated areas (large cities) by using low potency from transmitters already serving analog transmission integrated with small iso-frequency relays (gap-filler) to improve reception in critical areas. In the second phase, when the interference restrictions imposed by analog signals have been eliminated, it will be possible to transmit stronger digital signals so as to extend coverage to a greater percentage of the territory. The development of a digital television system through the air, however, though being a key objective, will take time and effort in the elaboration of legislation and norms governing the outset of these new services in different countries, and in the definition of a gradual transformation plan for the passage from analog to digital technology.

Television Broadcast via Satellite

From the operators' point of view, the satellite system offers greatest coverage possibilities for large geographic areas, giving access to vast catchment areas and overcoming those typical broadcasting difficulties of terrestrial broadcast systems.

Satellite, characterised by precise transmission standards, is the most common method used by the television theme channels included in packages from which the consumer can choose the channels they prefer.

Transmission is limited to homes in the coverage area with the necessary equipment for satellite reception (satellite dishes). Satellite systems, moreover, are limited to broadband transmissions. Narrowcasting operations, such as those used in interactive systems or telephony, cannot be provided by satellite.

The evolution of these Direct To Home (DTH) satellite systems will make them competitive with those initially forecast to be provided on the broadband terrestrial broadcasting infrastructure. The possibility of providing diffusive interactive services, such as Pay-Per-View and Near-Video-On-Demand (NVOD), acquires great relevance in this context thanks to the vast quantity of channels available in digital systems.

Naturally the success of these services among the public at large will depend on the terminal equipment costs. Terminal equipment is the set top box (or decoder) that sits on the TV and the antenna that goes with it. To this sum we must add the cost of subscription to the various services.

Today's digital broadcast satellites are superior to other methods of delivering television to the home in many respects. They offer more channels than are available over the air or from nearly any cable system. The quality of their pictures far exceeds that of off-air or cable broadcasters, even though their digital signals have to be converted into the old-fashioned standard NTSC[2] format for conventional TV sets. Table 4.1 summarises advantages and disadvantages of these systems.

Table 4.1: Advantages and Disadvantages of Direct To Home Systems

Advantage	Disadvantage
Great transmission capacity	Initial investment very high in terms of construction and launch of new satellites
Immediate availability	
Rapid service installation as cabling operations are not required	Complexity of any indirect interactive mechanisms (telephone network and satellite uplink)
Access infrastructures at users' expense (receiver and Set-Top-Box)	Service evolution potential limited by the assistance of a direct return channel.

Television Broadcast via Cable

Cable television has the greatest access to a large percentage of homes in United States and some countries in Europe (i.e., Germany). Cable television provides exactly the same services as over-the-air television insofar as any given program is concerned. Cable television provides features superior to over-the-air television in only two respects: it has far more channels, and it can provide two-way interactive services but requires appropriate investment in cabling and considerable installation time.

Despite this, many countries in Western Europe, with different time scales and methods, are planning the construction of widespread broadband cable infrastructures or, where they already exist, the modernisation of television cable networks.

The expansion of cable television is possible in densely populated areas, but is not economically viable in rural areas which represent 20-30% of the population in many European countries (in these areas microwave distribution, MMDS—*Multichannel Microwave Distribution System*—or *wireless* is often considered to be complementary to cable).

Interactive services seem to be a prerogative of cable systems where the *network narrowcast* can serve each user personally. Estimated costs can vary considerably depending on the context and the solutions adopted. Studies carried out in North American residential areas[3] show the overall cost of construction of an HFC (*Hybrid Fiber Coaxial*) network complemented with VOD (Video-On-Demand), telephony, and Internet access. The cost for each area for HFC network is $1,950,716 or $487,679 per home.[4] It should be borne in mind that this estimate involves almost exclusively above-ground cabling using pre-existing piling. Moreover, the addition of telephony to the service, where there is coaxial cable, involves the construction of a second superimposed network in copper duplex. The user apparatus is not included in this sum either.

Table 4.2: Advantages and Disadvantages of Television Broadcast via Cable

Advantages	Disadvantages
Possibility of transmitting both analog (CATV) and digital (DVB) signals.	Reduced upstream band: limits for interactive and telephony services.
The tree structure is efficient for broadcast services.	Complex mechanisms for interactivity and telephony (frequency allocation).
Consolidated technology.	High sensitivity to breakdown (tree structure).
High reliability of active elements.	Privacy problems on the coaxial section (telephony).
Immediate availability.	Does not support high-speed data services.
Direct interfacing with user apparatus (TV and STB).	

Television Broadcast via ADSL (Asymmetric Digital Subscriber Line)

The ADSL technique is a readily available solution for the supply of interactive multimedia services because it makes use of the pre-existing telephone network. It is therefore foreseeable that this technology will be used in the medium term to extend the supply of multimedia services to those areas not yet included in the new broadband infrastructure still under construction. In some areas, where territorial morphology and population density make neither cabling nor the implementation of a radio system economically viable, ADSL becomes the first choice. It is difficult, however, for ADSL to be used on a vast scale because of its limited growth potential in performance terms and the high number of apparatus necessary for each user connected to the service.

It should be noted that ADSL is a selective system: it allows for the use of one or two quality video programs at a time chosen from a much broader bouquet, therefore limiting the possibility of contemporary use from two or more terminals connected to the same line (as happens with traditional TV

Table 4.3: Digital Subscriber Line—xDSL

Comment	Speeds up to T1 (1.554 megabits per second)
Status	Experimental
Suitable for two way interactive video	Yes
Suitable for near video on demand NVOD	Yes but slow and low quality
Est. cost to serve 100 million U.S. household subscribers, each:	
- fixed cost of system	$1,000
- variable cost per household	$ 600
Capacity (TV channels available to each household with 5:1 digital compression	Less than one channel (VHS quality video)
Local content	Yes

Source: Adapted from Owen, 1999

within the family). Current advantages, disadvantages, and estimated costs of xDSL are summarised in Table 4.3.

Television Broadcast via Microwave MMDS (Wireless Cable)

Cellular television is an extremely interesting system because, though operating with microwaves at very high frequency, 40GHz, it is to all effects equivalent to cabling, which is rather expensive. In small cities or cities with a particular distribution, with small buildings, the use of this cellular television seems extremely interesting. While being a highly directive system, it is nonetheless a broadcasting system; it uses very high frequencies that are received by systems very similar to the satellite dish.

This system of transmission is of particular importance for partially satellite systems combined with local type channels or channels with local information, offering the possibility of making use of means of return connected to service centres of a local nature. This process involves cellular television that combines programs of national interest with local interest programs.

The principal attractions of the microwave system, from an economic point of view, lie in its low implementation costs and rapidity of plant installation. Recent comparisons with the CATV systems available today show that, assuming a 40% penetration, MMDS costs 50% less than an updating of existing coaxial cable infrastructure and 70% less that an analogous infrastructure in optic fibre.[5]

For high-frequency digital systems, international forecasts estimate that these proportions will remain unaltered with respect to cable solutions of equal performance. However, research carried out by the Secretariat of the Ministry for PT[6] in Italy came to different conclusions.

In particular two 40GHz infrastructure hypotheses were examined to cover the 21 regional capitals in Italy: one involving exclusively diffusive systems (Table 4.4.), the other for television services of the VOD type (Table 4.5.). These estimates did not include the costs of connection of cities to service centres or return connections from users to the same centres, nor did they include frequency concession licences.

This research reveals a good economic advantage when compared to a cabled solution of like performance, while interactivity would be strongly penalised in respect to a cable infrastructure. Indeed in this case, in return for a slightly inferior cost on a cabled solution, the resulting service would be of little

Table 4.4: Estimated Cost of Microwave Diffusion System

No. Channels	128, flows at 4 Mbit/s
No. Users served	15 billion, 4,5 thousand families
Network cost	$ 145 million
Cost per inhabitant	$ 10
Cost per family	$ 36

Source: Ministry PT Italy

Table 4.5: Estimated Cost of Microwave VOD System

Concentration	1 channel every 5 users
No. Users per cell	2.500
No. Overall users	1 thousand
Network cost	$ 568 million
SDH Radio bridge at 155 Mbit/s per cell interconnection	$ 191 million
Cost per inhabitant	$ 0,77
Cost per family	$2,71
Average cost of one programme in each capital	$ 284

Source: Ministry PT Italy

interest in terms of potential users while using a significant portion of the available frequency spectrum.

Though it is necessary to examine the estimates more in depth on the basis of future real market offer, in general the clearest advantage of a radio system when compared to the cable alternative lies in the flexibility and speed of structural construction. A radio system does not require the impressive public works involved in laying cables, an important issue in countries such as Italy with many cities of great historical significance. A radio system, moreover, can be easily restructured and expanded. Such features drastically cut the installation costs and implementation times: while the cabling of vast areas would take years, a radio system can become operative within a few months.

Such systems, however, have some limits that reduce their general opportunities of use such as the lack of amplitude of the cells covered, making use in scarcely populated areas improbable; ineffectiveness in mountainous regions, and sensitivity to meteorological phenomena, in extreme cases can cause a complete interruption in the service.

Table 4.6: Advantages and Disadvantages of Microwave Access Systems

Advantages	Disadvantages
Low costs (estimated) Speedy Implementation Little urban impact Noteworthy development potential (LMDS)	Sensitivity to the geographic conformation of the territory Lack of uniformity in international norms Difficult to include in the national frequency scheme Complex user apparatus Sensitivity to meteorological conditions

Source: Reseau, 1997

Comparative Analysis of Different Transmission Technologies

Table 4.7. shows a synthetic table of the advantages and limits of different transmission technologies for digital television signal.

Table 4.7: Advantages and Limits of the Different Broadcast Methods

	Satellite	Terrestrial	Cable	ADSL	MMDS
Number of digital channels	200+ channels	20-30 channels	100 channels but 400+ with the appropriate cables	All signals received on demand without limit in the number of channels	about 60 channels
Audience	All homes in the area covered equipped to receive the signal	Almost all the homes in the state will be able to receive the signal	Only homes reached by the cable can receive the signal (extension of network coverage)	All homes with a telephone line can receive the signal	All homes with an antenna within coverage range
Reception apparatus	Satellite dish and set-top-box necessary	Antenna Set top box necessary	Connection to the cable network Set top box	Modem and set top box	Antenna on the roof of the building and set top box
Area Covered	National and multinational transmissions	National and regional transmissions	Regional and local transmissions	Programmes available globally	Range of cover in 10s of km and system only aimed at providing broadcasting services alternative to a local CATV infrastructure
Interactive services	Back channel through the land telephone line (narrowband)	Back channel through the land telephone line (narrowband)	Back channel incorporated in the system. Broad bandwidth	Interactive service possible by duplex without any investment in optic fibre network or coaxial cable.	Limited interactive services pay TV or NVOD associated to a method of indirect interaction or telephone network
Transmission infrastructure	Relatively inexpensive infrastructure	Inexpensive infrastructure	Expensive infrastructure; cabling necessary	Infrastructure already exists but back up signal requires added investments	Inexpensive infrastructure

In digital transmission, satellite and cable offer important advantages, but are more limited when compared to the terrestrial network. The advantage of satellite is quantitative and consists in the number of television channels broadcast, which is considerably higher than those available on the terrestrial network. This depends on terrestrial frequency allocation, which varies from country to country.

Cable systems have advantages that are not only quantitative (they also carry more digital channels than the terrestrial network) but also qualitative: to the transmission of television channels, they can add broadband interactive services.

Finally, ADSL technology (*Asymmetric Digital Subscriber Loop*) offers the greatest degree of interactivity.

The advantages of a microwave system, MMDS, over the cable alternative exist in the flexibility and speed of infrastructure construction.

Which system is cheapest? The answer depends on which set of services or capabilities is the minimum required to attract consumers.

Terrestrial wireless systems—such as a Local Multichannel Distribution Service (LMDS)—are extremely appealing, chiefly because of their low fixed costs. These wireless systems do not require massive up-front investments which lowers the financial risks considerably.

All of the video delivery systems have one thing in common: they cost more than the average household is now willing to pay for video services, even video services integrated with Internet content.[7]

Comparing the cable network with satellite, it is possible moreover to underline some strengths and weaknesses in both systems.

From the operators' point of view, the satellite system offers the greatest coverage possibilities for vast geographic areas, thus giving access to vast catchment areas and overcoming the typical problems of terrestrial transmission.

Cable operators instead must lay cables across the territory in order to reach their users. Therefore their networks tend to be concentrated in densely populated areas, excluding rural areas with less coverage.

Moreover, satellite systems are limited to broadband transmissions. Narrowcast operations, such as services with an high degree of interactivity and telephony, cannot be supplied by satellite. Consequently real interactivity appears to be a prerogative of cable systems where narrowcast networks can be used to serve each user individually. These differences are shown in Table 4.8.

Table 4.8: Principal Advantages and Disadvantages of Transmission Methods via Cable and Satellite

	Cable	Satellite
Audience	Limited to network coverage range	National/multinational potentially global
Interactive services	Simple to introduce	More difficult to introduce because of the need to use telephone land lines
Telephony services	The network can supply them. Implementation can be easy depending on network quality	Not possible
Coverage	Rural areas are low priority because of low population density	No problems
Investment per home	Proportional to number of homes reached (in urban areas)	Fixed cost in satellite investment
Personalised services	Relatively easy to implement	Cannot be introduced

We can compare capitalized costs of satellite versus cable systems (see Table 4.9) in order to evaluate which transmission system is cheaper per household.

A system such as Hughes's DSS (Table 4.9) costs $750 million or more to launch, and each subscriber requires a set top box that costs at least $200. Nevertheless, all it takes is 10 million subscribers to bring the capital cost per subscriber down to $275, and 10 million subscribers does not begin to exhaust the potential market of 100 million TV households in the United States.

A large urban cable system has a construction cost of at least $600 per subscriber, maintenance expenses of about $60 per subscriber per year, and drop lines and set top boxes that total another $75 or more per subscriber. Such a system might have 100,000 subscribers and therefore an overall capitalized cost per subscriber of $1,275. Not only is this number much higher than the satellite cost, but it cannot be reduced significantly through subscriber growth,

Table 4.9: Capitalized Costs of Satellite vs. Cable Systems

	Satellite	Cable
Number of channels	200	60
Number of subscribers	10,000,000	100,000
Cost of fixed upfront investment	$750,000,000	$60,000,000
Subscriber equipment cost	$200	$75
Plant maintenance cost	0	$60/yr, or $600 capitalized
Programming	Network	Local and network
Capital cost per subscriber	$275	$1,275
Capital cost per subscriber per channel	$1.38	$21.75

Source: Owen, 1999

because on average 70% of all U.S. households already subscribe to cable service. The satellite cost advantage is even more striking on a per-channel basis.

Compared with any terrestrial system, wired or wireless, a satellite system has economic leverage because the incremental cost of serving a subscriber anywhere in the United States is close to zero, on-premises equipment excluded. Both wired and wireless broadcast systems must build new common facilities in order to extend their geographic reach or to add digital channel capacity. The satellite advantage thus becomes even greater if on-premises customer receiving equipment costs are minimized, perhaps by building much of it directly into digital TV sets.

The chief marketing disadvantage for DBS today is the difficulty in getting local programming. Television sets equipped with satellite receivers must also be connected to something else (rabbit ears, cable, or rooftop antenna) in order to receive local TV stations. DBS providers have tried to minimize this inconvenience by making available on the satellite a package of broadcast network signals (ABC, NBC, CBS, Fox, PBS). These signals are picked up from one or more affiliates of each network and rebroadcast under a statutory compulsory copyright license intended to provide service to rural areas. In any event, this service does not provide local programming such as news and hometowns sports. Many Ku-band DBS subscribers apparently continue to subscribe to basic cable service in order to receive local stations, adding around $16 per month to the cost of satellite service.

THE ECONOMIC EFFECTS OF DIGITAL TRANSMISSION

The change-over to digital television transmission (via airwaves, cable, satellite, ADSL, and MMDS) has different economic effects on some of the different categories involved: state, users, providers (distribution systems). Below is an analysis of the principal economic implications for each of the categories.

The State

The state has an important role in the decision to change over to digital terrestrial transmission. First it can create the conditions for more efficient use of the frequency spectrum, allowing for the disengagement of part of these frequencies, which, depending on the choices made, can be allocated to other terrestrial television channels, other forms of transmission (data broadcasting), telecommunications (interactive mobile systems), or can be divided between these different functions.

The state has the possibility of rationalising the use of a precious national resource (the electromagnetic spectrum), creating space for new economic activity and collecting resources for the rental or sale (as in the USA) of a part of the spectrum.

For evaluation and recompense of the spectrum, in radio and television, almost all European governments have chosen the option of delegating (by concessions or authorisations) the necessary frequencies through administrative procedures established and regulated nationally by the Ministry for Communications or independent regulatory bodies. On the basis of established parameters, such procedures concede the exclusive use of a part of the frequency spectrum to a licence holder for a given service within the territory for a period of time in exchange for a specified fee.

The alternative option would be the auction of the frequencies for a specific service. In this case it would be the economic operators who would determine the economic value of the frequencies with their offers, bearing in mind the established limits (duration and means of rights transfer, etc.). This auction method was used in the USA when the government sold the frequencies for personal telecommunications services (PCS) with a revenue of about $10 billion ($38 per inhabitant). This second option was partially adopted in the UK in 1992 during the renewal of regional Independent Television concessions.

In conclusion, digital television offers some advantages to governments, such as:

- the proceeds from the "sale" of frequencies;
- a rapid development of digital markets both in interactive television services and mobile communications;
- an international first mover advantage in the digital television sphere.

The Viewer

The economic effects of the passage to digital television transmission for the viewer are to be seen in the costs of digital reception. These costs vary according to transmission technology.

Table 4.10: Comparative Analysis of Costs for the Viewer in Cable Television Reception

Country	Channel	Cost of decoder	Installation	Pricing
France	NC Numericable	Rental: € 7,47 monthly plus € 76,25 returnable deposit	€ 106,76	Per month from € 6,87 for aminimal channel budget up to € 16,88 for all channels
France	Noos (Lyonnaise Cable)	Rental: € 6,87 monthly plus € 76,25 returnable deposit of	€ 106,76	Per month from € 4,57 for a minimal channel budget up to € 54,90 for all channels
France	France Telecom Cable	Rental: € 6,87 monthly plus € 76,38 returnable deposit	€106,76	Per month first bouquet € 10,67. Second, third and fourth € 7,62 each. With four bouquets, the fifth and sixth are free.
Germany	Media Vision	Purchase € 409 Rental: € 7,59 monthly plus € 76,60 returnable deposit	n.a.	Per month from € 5 for a minimal channel budget up to € 12,70 for all channels
Spain	ONO	Rental: € 23,29	€ 59,73 rental connection p/m € 4,65	Per month from € 6,67 for a minimal channel budget up to € 21,35 for all channels
UK	Telewest	n.a.	n.a.	Per month from € 17,82 for a minimal channel budget up to € 32,64 for all channels
UK	NTL	Rental: €16,28 monthly	€ 64,80	Per month from € 22,68 for a minimal channel budget up to € 42,12 for all channels
UK	C&W Communications	Rental: £ 9.96 per month	£50	from £12,98 up to £28,45
Sweden	Comhen	Purchase: € 226-354 or included in subscription package	No cost except for the smartcard rental to activate the service at € 27.65 for the first year and € 21.97yearly for the following years.	Per month from € 13,79 for a minimal channel budget up to € 35,26 for all channels

Source: Elaboration of New TV Strategies Database, 2002

In the case of cable transmission, the costs lie in the initial cabling connection of the home to the service provider and subsequently the cost of a monthly subscription. Digital transmission via cable also involves the cabling within the home when a second or third television set is added.

Satellite transmission requires special terminal equipment (satellite dish, set top box) and the purchase of a program package from the satellite program provider. In reference to the single user, the principal cost categories are:

- purchase of the satellite dish;
- installation costs;
- purchase or rental of the digital receiver;
- subscription to a bouquet of channels.

To receive digital transmissions via terrestrial network, the viewer must also have a set box top or decoder, and in the majority of cases it is sufficient to have the normal antenna necessary for analog television. The number of households that have to change antenna will depend on the frequency-planning scheme and technical factors that vary from country to country.

Table 4.11: Comparative Analysis of Costs for the Viewer for Satellite Reception (satellite dish excluded)

Country	Channel	Cost of digital decoder	Installation cost
UK	Sky Digital	Free	€65,12 for standard installation, subject to 12-month minimum subscription agreement. Otherwise €162,8
France	Canal Satellite Numerique (CSN)	Rental: € 6,86 per month plus € 76,25 returnable deposit	€38,12 connection fee
France	Television Par Satellite	Rental: € 6.87 per month plus € 76,26 returnable deposit Purchase: € 455,87	€ 38.13
France	ABSat	Rental: € 6.86 per month	€ 38.13 connection fee plus € 15,25 for smart card
Germany	Premiere World	Rental: € 7,60 per month plus € 76,67 deposit	€ 15,27
Spain	Canal Satellite Digital (CSD)	Rental: € 5,99 per month	n.a.
Spain	Via Digital	Rental: € 7,19 per month	n.a.
Italy	Tele+ Digitale	Rental € 7.28 per month plus € 51,24 deposit	€ 46,28 smart card
Italy	Stream	Rental € 7.23 per month	n.a.
Sweden	Canal Digital	Rental: € 8,15 per month purchase: € 591	n.a.

Source: Elaboration of New TV Strategies Database, 2002

Table 4.12: Price Dynamics for Advanced Internet Access Systems in Relation to "Affordability Level" (access threshold for consumers)

Systems	Affordability Level ($)	Average annual price variation from 1990 to 1999	Average annual price variation from 1999 to 2003	Year in which Affordability level will be reached
Optic fibre cabling	500	-15%	-15%	after 2003
ADSL	500	-24%	-18%	1999
Cable Modem	500	-23%	-21%	2000
DVB Box	400	-19%	-16%	1997
ISDN	200	-16%	-19%	1998

Source: Elaboration of Databank Consulting Data by Andersen Consulting, Booz Allen & Hamilton, Dataquest, Ovum, Telecommunications, Telephony, Yankee Group, 1997

In the United Kingdom, where terrestrial digital television was launched in 1998, it is estimated that only a small percentage of households will have to purchase a new antenna, alongside the set top box and a low-cost domestic antenna (less than $15).[8] Also in the case where a viewer needs a new antenna, the relative costs are lower than those for the installation of cable and satellite.

Table 4.12 shows the Affordability Level or the price threshold within which the viewer can afford to purchase the different multimedia systems including ADSL and Cable Modem. The table also indicates the year in which this threshold may be reached for each different system.

The Operators Involved

The possible winners in the conversion to digital are therefore:

- *the Treasury*: with the revenue from auctioning of available frequencies;
- *the average family* without Pay TV to date, passing from five to over 10 uncoded channels and many more interactive services and digital channels on payment;
- *industry which manufactures and sells TV sets* which seems destined to undergo accelerated growth;
- *advertising agencies* which receive a budget (not less than £100 million in the UK) for marketing the launch of digital TV;
- *telecommunications operators* who will have an increase in the use of telephone networks thanks to interactive services;
- *content providers* who will have a new and profitable distribution network;
- *the owners of transmission infrastructure* who will have numerous new clients.

Not all of these benefits will bring net gains; in many cases it is a matter of transfer from one category of citizen to another. The consumer will have to bear the greater part of the expense, such as:

- purchase of a new digital television or decoder for digital reception;
- purchase of a new antenna in the case that the frequencies received from the multiplex are not from an analog relay;
- in some cases the purchase of a more modern antenna to ensure a stronger signal.

The function of the video recorder will be reduced with digital transmission, as it will no longer be possible to record a different television channel from the one being watched.

Moreover the *multiplexers* will have to provide a toll-free number for assistance to families that do not receive the signal well and require a new antenna.

As consumers will contribute a substantial share of the necessary investment for digital conversion, it is essential that the benefits of this new technology are apparent to the average family. To date only 30% of British families have shown active interest in Pay TV (via cable or satellite), and the others express satisfaction with the present five uncoded channels available.[9] Digital operators will have to convince them through special discounts, new programs, and new interactive services.

COSTS AND PENETRATION TIMES COMPARED

The costs of the three modes of digital television transmission (cable, satellite, terrestrial broadcast) can be compared by means of a mathematical model aimed at estimating the penetration prospects of each one.[10] The reference context is a European country characterised by the following significant features:

- 20 million families in possession of a television;
- a volume of frequencies reserved for land broadcast of digital television equal to four analog channels;
- six digital channels substituting each analog channel;
- 25% of the land channels with a regional window;
- 30% of homes obliged to install a new antenna for reception of digital land broadcast;
- set top box, for terrestrial and digital cable, initially priced at € 450 which should drop to € 120 after 12 years;
- set top box, for satellite and digital cable, initially priced at € 410 which should drop to € 110 after 12 years.

The principal cost of the three television systems examined lies in transmission, distribution, and signal reception. Though relevant, the costs of program production will not be considered since they do not vary according to the system adopted.

- Distribution costs are those required to send the signal to the transmitter (for land), the transponder (satellite), or the head-end station for cable.
- Transmission costs are those involved in the diffusion of television services from the distribution point to the viewers. The estimate of necessary investments for the creation of infrastructures is based on average European market prices and bears in mind that digital distribution systems can often be built using existing analog infrastructures.
- Reception costs are those paid by the family to receive digital signals on the main domestic television set. The price of the set top box is that declared by the manufacturing industry.

In the case of a **digital terrestrial** network, it can be assumed that it will use the existing national infrastructure, as happens in many European countries. Increases in costs only involve updating the existing broadcast network, construction of new antennae on masts, and the installation of new transmitters where necessary (costs relative to the installation of sites and transmission masts are excluded), presuming that only a small portion of antennae will require substitution or re-orientation.

It must be borne in mind that the network will have to offer the option of creating a certain percentage of regional broadcasting.

In the case of **digital satellite,** the study presumes that a new satellite is necessary for the service; indeed the satellites currently in use for analog transmission have reached a broad base of subscriptions that cannot be automatically transferred to another medium.

Cost forecasts include the acquisition and launch of a satellite equipped for direct digital broadcast, the construction of plants for the transmission up-link segment (earth-satellite), and a series of other elements typical of satellite transmission (e.g., a back-up to be used in case of breakdown).

In reference to important differences in cost between national and regional coverage, the study covers two different scenarios:

- 12 programs with national coverage and 12 with regional coverage;
- 24 programs covering the national territory homogenously.

In the case of **digital cable**, considering the great difference of penetration rate among the European countries, the real costs are based on the following two hypotheses:

- the cost of constructing digital cable networks (where they do not exist at present);
- the cost of converting analog cable networks to digital.

The elaboration of the above data has given the results shown in Figures 4.1 and 4.2.

Figure 4.1 shows the total costs of the system, in other words the costs to broadcasters and consumers in relation to:

- distribution (transfer of the signal from the broadcaster to the point of on-network transmission);
- transmission (from the input point to the user's home);
- equipment necessary for the user in order to receive or view programs transmitted digitally.

Figure 4.2 instead shows the distribution and transmission costs and only includes investments made by the broadcaster.

By comparing the two figures, it is clear that the cost of the set top box (and the necessary extra apparatus for reception) is the most important component of the system's overall cost. The greater part of the overall cost is therefore covered by the viewer.

Figure 4.1: Costs for the Entire Television System

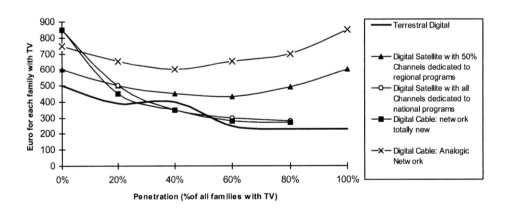

Source: Elaborated from Data of Convergent Decisions Group, 1998

Figure 4.2: Costs of Distribution and Transmission

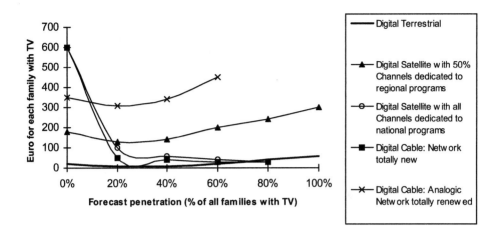

Source: Elaborated from Data of Convergent Decisions Group, 1998

An analysis of Figure 4.1 reveals that the digital terrestrial transmission system is the least costly means of reaching total penetration on a national scale.

According to this hypothesis (Figure 4.2), the costs incurred by the broadcaster are very low, up to a penetration of 50% of households. Costs tend to increase above this threshold in relation to the greater number of transmitters and relays needed to cover the more difficult areas to reach (peripheral areas or areas with reception difficulties).

According to the estimates shown,[11] a transmission network that covers 90% of the population can be created at low costs within a period of two to five years if existing transmission sites are used. To reach the coverage of the present analog network, which exceeds 99% of the population, would involve much more time and strongly increase marginal costs.

The costs necessary to reach the 10% of the population that resides in remote areas are high in relation to the possible profits.

In the case of digital satellite, the total cost per household is very high, also in correspondence to the low level of penetration. This means of transmission necessitates high fixed costs for launching the service, which in contrast to the other methods of transmission are independent of the number of homes served. Figures 4.1 and 4.2 show two different hypotheses related to satellite: the first option, the more costly, includes a quota of channels for regional programming; the second only involves national channels.

Satellite is at its most cost advantageous when it involves total coverage of the national territory. It is important to note (as emerged from the section above) that the apparatus necessary for satellite reception (satellite dish and relative equipment) is more costly than the apparatus for the other systems.

The cable mode of digital transmission has more expensive cost levels than the hypotheses of terrestrial or satellite transmission. It is possible, moreover, to note different levels of cost in the conversion of analog networks to digital and the construction of a completely new digital network. The investments necessary to lay cable in densely populated areas can de divided among a substantial number of users, but outside of urban areas, marginal costs increase progressively.

From the data in Figures 4.1 and 4.2, it becomes clear that the most cost-effective system to supply digital television to almost all the families in Europe is the digital terrestrial broadcast system. If the level of penetration should surpass 98%-99%, digital satellite transmission could compete with digital terrestrial broadcast. In terms of cost, in countries such as Italy, this threshold is slightly lower given the particular population concentration and the widespread presence of mountainous regions.

ENDNOTES

[1] Richeri, G. (1999). Reti di trasmissione digitali terrestri: Una strategia di lungo periodo. In *La Televisione Digitale Terrestre: Tendenze di Sviluppo, Vantaggi e Problemi*. Milan, Italy: Quaderni di Economia dei Media. Fondazione Rosselli.

[2] NTSC—National Television Standards Committee. Used to refer to the technical standards and physical characteristics of conventional analog TV broadcasting, as enshrined in FCC regulations.

[3] Territorial context: 100,000 inhabited one-family units, over an area of 95km,^{2} subdivided in areas of 4,000 homes each. Source: Dittberner Associates, 1995.

[4] Source: Reseau, 1997.

[5] Source: Reseau, 1997.

[6] Report of the workgroup. (1996). *Study on the Application Possibilities of MMDS/LMDS Technologies to Microwave Distribution of Television Services, and More Generally, New Multimedia Services*. Rome, Italy: Ministry for PT Italy.

[7] Owen, B.M. (1999). *The Internet Challenge to Television.* Cambridge, MA: Harvard University Press, p. 331.

[8] Source: Davenport, H. (1998). Better than expected "coverage of ONDigital. *New Media Market,* 16(43).

[9] Murroni, C. (1999). Gli sviluppi nel Regno Unito. In *La Televisione Digitale Terrestre: Tendenze di Sviluppo, Vantaggi Problemi.* Milan, Italy: Quaderni Istituto Economia dei Media, No. 9, Fondazione Rosselli, pp. 32-43.

[10] Source: Convergent Decision Group. (1998). *Digital Terrestrial Television in Europe.* London: Convergent Decision Group Report.

[11] Source: Convergent Decision Group. (1998). *Digital Terrestrial Television in Europe.* London: Convergent Decision Group Report.

Chapter V

Interactive Digital Television*

INTRODUCTION

Interactive television (iTV) can be defined as the result of the process of convergence between television and the new interactive digital technologies (Pagani, 2000).

Interactive television is basically domestic television boosted by interactive functions that are usually supplied through a "back channel" and/or a modern terminal. The distinctive feature of interactive television is the possibility that the new digital technologies give the user[1] the opportunity to interact with the content that is offered.

The evolution towards interactive television has not an exclusively technological, but also a profound impact on the whole economic system of digital broadcasting—from offer types to consumption modes, and from technological and productive structures to business models.

This chapter attempts to analyse how the addition of interactivity to television brings fundamental changes to the broadcasting industry.

---

* Earlier versions of this paper were presented at the 6th Symposium on Emerging Electronic Markets, at the University of Muenster (Germany), September 1999; the Business Information Technology Conference at the Universidad Iboamericana in Mexico City—Santa Fé, May 2000; and the SISEI Conference at Bocconi University in Milan (Italy), February 2001. The author wishes to acknowledge the input of participants at these conferences.

The chapter first defines interactive transmission systems and classifies the different services offered according to the level of interactivity determined by two fundamental factors such as response time and return channel band.

After defining the conceptual *framework* and the technological dimension of the phenomenon, the chapter considers the impact generated by interactive digital technologies on the whole broadcaster economic system.

Some dimensions of the economic sub-system of reference are considered such as new types of interactive services offered and new competitive system emerging.

The Interactive Digital Television (iDTV) value chain will be discussed to give an understanding of the different business elements involved.

The results of the present section allow for an understanding of the impact of interactive television on the whole economic system, together with the significant changes in the market, operator types, and distributive systems. There are many problems that management has to deal with as a result of the changing behaviour of the audience, the status of the viewer, and the nature of the medium and its function. Technological, organisational, and service innovation is undoubtedly the key for understanding the behaviour of firms and institutions in the development of interactive television.

A DEFINITION OF INTERACTIVITY

The term interactivity is usually taken to mean the chance for interactive communication among subjects. Technically, interactivity implies the presence of a return channel in the communication system, going from the user to the source of information. The channel is a vehicle for the data bytes that represent the choices or reactions of the user (input).

This definition classifies systems according to whether they are diffusive or interactive.

- Diffusive systems are those which only have one channel that runs from the information source to the user (this is known as *downstream*).
- Interactive systems have a return channel from the user to the information source (this is known as *upstream*).

There are two fundamental factors determining performance in terms of system interactivity. These are response time and return channel band.

Table 5.1: The Classification of Communication Systems

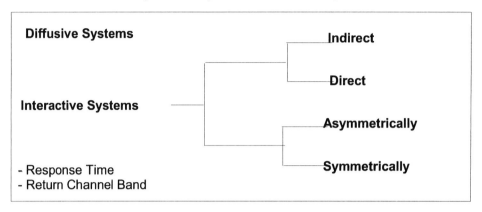

The more rapidly a system's response time to the user's actions, the greater is the system's interactivity. Systems can thus be classified into:

- indirect interactivity systems, when the response time generates an appreciable lag from the user's viewpoint;
- direct interactive systems, when the response time is either very short (a matter of a few seconds) or is imperceptible (real time).

The nature of the interaction is determined by the bit-rate that is available in the return channel. This can allow for the transfer of simple impulses (yes-no logic), or it can be the vehicle for complex multimedia information (as, for example, in the case of videoconferencing). From this point of view, systems can be defined as asymmetrically interactive when the flow of information is predominantly downstream. They can also be defined as symmetrical when the flow of information is equally distributed in the two directions.

Based on the classification of transmission systems above adopted, multimedia services[2] can be classified into diffusive (analog or digital) and interactive (Table 5.2).

Interactive multimedia services include both diffusive numerical services (Pay-Per-View, Near-Video-On-Demand) and asymmetrical interactive video or up-to-date types of symmetrical interactive video. Digital television can provide only the first two categories of multimedia services: diffusive numerical services and asymmetrical interactive video services. Services such as videoconferencing, telework, and telemedicine, which are within the symmetrical interactive video based upon the above classification, are not part of the digital television offers.

Table 5.2: Classes of Service (classes not directly relevant to interactive multimedia services are in gray)

Class of services	Services (examples)
Analog transmission	Free channels Pay TV Pay-Per-View (PPV)
Numerical diffusion	Pay Per View Near-Video-On-Demand (NVOD)
Asymmetric interactive video	Video-On-Demand (VOD) Music-On-Demand Home shopping Video games Tele-teaching Access to data banks
Low-speed data	Telephony (POTS), data at 14,4; 28,8; 64; 128 Kbit/s
Symmetric interactive video	Cooperative work Tele-work Tele-medicine Videoconference Multi-videoconference
High-speed data	Virtual reality, distribution of real-time applications

Source: Adapted from Reseau and Intel

The different types of interactive multimedia services described can be classified into categories based upon their level of interactivity attributed to each one of them, based upon a 0 to 4 score (Table 5.3).

Table 5.3: Levels of Video Interactivity

Level	Description	Applications	Forward Channel	Return channel
0	Call to the service provider by the viewer	Pay per view, pay Tv	Broadcast Network	Usually telephone
1	Pseudo-interactivity	Teletext, Internet access, games		Information locally stored on Tv
2	Basic interactivity from remote control	PPV, VOD, home commerce, Internet		Wireless (es. GSM, DECT, UMTS) is plugfree and doesn't require professional installation
				Integrated return path: suitable for cable the antenna can be used for the back signal
3	Use of return channel for video reception	Video on demand; Internet video streaming, Videotelephone, two way video	Switched network	Telephone up-to-date with:ISDN/ADSL/UMTS
4	Network with full service	Two way Video; professional use i.e. telemedicine, videoconferencing		Modem cable; VSAT (satellite)

Local Interactivity

An interactive application that is based on local interactivity is one that does not require a return-path back to the service provider. A good example is the broadcaster transmitting a football match using a "multi-camera angle" feature, transmitting the video signals from six match cameras simultaneously in adjacent channels. This allows the viewer to watch the match from a succession of different vantage points, personalising the experience.

One or more of the channels can be broadcast within a time delay, for instant replays. As far as the viewer is concerned this experience is unique to him or her, but it involves no signal being sent back to the broadcaster to obtain the extra data: the information is already there, being broadcast in a linear fashion in much the same way as the video signal is. The viewer is simply dipping in and out of that datastream to pick up supplemental information as required.

Technically, of course, there is a return path operating in all such applications, but it is between the user's remote control and the TV set, and goes no further.

That is why this type of application is described as exhibiting local interactivity. In a digital TV context, these are commonly referred to as "enhanced TV" applications.

One-Way Interactivity

It would be possible to offer an interactive application in which the viewer did send back a signal to the service provider via a return path of some kind, but without the intention (or even the ability) of invoking a direct response at that time. For instance, TV viewers could be "polled" by asking them to press a certain button on their remote control at a certain time to express a particular opinion—perhaps during an election. Alternatively, viewers could "play along" with a TV game show, sending their responses back to the service provider. The winning entry could result in a prize being dispatched to the appropriate viewer.

The most obvious application, however, is direct response advertising. The viewer clicks on an icon during a TV commercial if interested in the product, which sends a capsule of information containing the viewer's details to the advertiser, allowing a brochure or sample to be delivered to his/her home. There are obvious analogies to this in the PC and mobile telephony environments.

This can be defined as one-way interactivity. Messages can be sent back to the service provider, but there is no genuine, ongoing, continuous, two-way, real-time dialogue, and the user doesn't receive a personalised response. It allows a form of communication, although not a sophisticated one.

Two-Way Interactivity

Two-way interactivity is what the technological purist defines as "true" interactivity. The user sends data to a service provider or other user which travels along a return path, and the service provider or user sends data back—either via the return path itself or "over the air." Two-way interactivity presupposes "addressability"—the senders and receivers must be able to address a specific data set to another sender or receiver and to no one else.

What might be termed "low-level" two-way interactivity is demonstrated by a TV Pay-Per-View service. Using the remote control, the viewer calls up through an on-screen menu a specific movie or event scheduled for a given time and "orders" it. The service provider then ensures—by sending back a message to the viewer's set top box—that the specific channel carrying the movie at the time specified is unscrambled by that particular box, and that that particular viewer is billed for it. This is "low-level" interactivity in the sense that hardly any data at all is being passed back to the service provider, although there is a real-time "result." In the PC and mobile environment, any pay-per-use application would be an illustration of this. Low-level two-way interactivity is characterised by the fact that the use of the return path back to the service provider is peripheral to the main event.

"High-level" two-way interactivity, on the other hand, is characterised by a continuing two-way exchange of data between the user and the service provider, an exchange that is fundamental to the application. A good example of this type of interactivity is represented by videoconferencing—whether on a PC or (in a few years) on a mobile. Other obvious examples are Web surfing and multiplayer gaming, and communications-based applications such as chat and SMS messaging.

TV applications which either offer one-way or two-way interactivity tend to be referred to generically as "interactive TV" applications as opposed to "enhanced TV" applications.

"INTERACTIVITY": PROTOTYPE, CRITERIA, OR CONTINUUM?

Taking a look at the collection of existing definitions of "interactivity" spread throughout media studies and computer science, it seems that there are three principle ways of defining the concept:

1) as prototypic examples;
2) as criteria, i.e., given features or characteristics that must be fulfilled;
3) as a continuum, i.e., as a quality which can be present to a greater or lesser degree.

A definition by *prototypic example* is suggested by Durlak[3] (1987) who affirms that interactive media systems include the telephone, "two-way television," audioconferencing systems, computers used for communication, electronic mail, videotext, and a variety of technologies that are used to exchange information in the form of still images, line drawings, and data.

This type of definition is never, by its very nature, very informative, partly because it doesn't point out which traits qualify a given media as interactive or which aspects connect them. This concept of "interactivity" refers both to media patterns of the consultational and the conversational type. It is related to the sociological concept of "interaction" (in the form of the conversational communication pattern) and borrows from the informatic concept of interaction (in the form of the consultation communication pattern).

Examples of the second type of definition—*interactivity defined as criteria* is represented by Carey[4] (1989), who suggests that interactive media include all technologies that provide person-to-person communications mediated by a telecommunications channel (e.g., a telephone call) and person-to-machine interactions that simulate an interpersonal exchange (e.g., an electronic banking transaction).

Most of the content is created by a centralized production group or organization, and individual users interact with content created by an organization.

This conceptual construction points directly toward the conversational media type and the consultational media type, respectively, (and as a result, at the sociological and informatic concepts of interaction) which collectively make up "interactive media." Once again there is a certain vagueness to the definition of the concept, and the definition excludes services based on the transmission pattern, such as teletext, datacasting, Near-Video-On-Demand, etc., which make up the bulk of some TV systems' so-called "interactive services."

The third possibility, which solves some of these problems, is to define interactivity not as criteria, but rather *as a continuum*, where interactivity can be present in varying degrees. One possible way to structure this type of definition is to base it on the number of dimensions it includes, so that we could speak of one-dimensional, two-dimensional, three-dimensional... and n-dimensional interactivity concepts.

Rogers[5] (1986) suggests a relatively simple model of interactivity as a continuum, which operates from only one dimension, and he defines "interactivity" as the capability of new communication systems (usually containing a computer as one component) to "talk back" to the user, almost like an individual participating in a conversation. Rogers considers interactivity as a variable; some communication technologies are relatively low in their degree of interactivity (for example, network television), while others (such as computer bulletin boards) are more highly interactive.

Based on this definition, Rogers creates a scale, in which he lists "degrees of interactivity" for a number of selected communication technologies on a continuum from "low" to "high."

The basic model is clearly "human-machine interaction," understood in the context of interpersonal communication ("talking back"). It is also because of this consultational aspect (selection available between channels and programs) that classical transmission mass media, such as TV, can be considered "interactive"—although to a lesser degree. As is apparent, this attempt to sort and define is relatively rough and lacking in information.

Szuprowicz,[6] among others, has presented a two-dimensional concept of interactivity. Interactivity is defined by the type of multimedia information flows, divided into three main categories (Szuprowicz, 1995):

1) "User-to-documents" interactivity is defined as traditional transactions between a user and specific documents, and is characterized by being quite restricted since it limits itself to the user's choice of information and selection of the time of access to the information.
2) "User-to-computer" interactivity is defined as more exploratory interactions between a user and various delivery platforms characterized by more advanced forms of interactivity which give the user a broader range of active choices, including access to tools that can manipulate existing material.
3) Finally, "user-to-user" interactivity is defined as collaborative transactions between two or more users, in other words, information flows which make direct communication between two or more users possible. This last

form, contrary to the first two mentioned above, is characterized, among other things, by operating in *real time*.

Where the first dimension in the matrix is made up of these various information flows, the other is made up of other aspects, which these flows are dependent upon, here again divided into three categories: "access, distribution, and manipulation of multimedia content."

Laurel[7] gives an example of three-dimensional concepts of "interactivity," arguing that interactivity exists on a continuum that could be characterized by three variables, specifically (Laurel, 1991):

1) "frequency," in other words, "how often you could interact";
2) "range," or "how many choices were available";
3) "significance," or "how much the choices really affected matters."

Judged by these criteria, a low degree of interactivity can be characterized by the fact that the user seldom can or must act, having only a few choices available that make only a slight difference in the overall outcome of things. On the other hand, a high degree of interactivity is characterized by the user having the frequent ability to act, having many choices to choose from.

Understood in this way, the concept can be said to point out three aspects of "interactivity" within the consultation communication pattern.

An example of a four-dimensional concept of interactivity can be found in the writing of Goertz,[8] who simultaneously presents a considerably more elaborate attempt at a definition, and he isolates four dimensions, which are said to be meaningful for "interactivity"(Goertz, 1995):

1) the degree of choices available;
2) the degree of modifiability;
3) the quantitative number of the selections and modifications available;
4) the degree of linearity or nonlinearity.

Each of these four dimensions also makes up its own continuum, which Goertz places on a scale. The higher the scale value, the greater the interactivity.

Finally, there are concepts of interactivity which operate with more than four dimensions, e.g., Heeter's[9] six-dimensional concept of interactivity (Heeter, 1989).

A definition as a continuum appears to be more appropriate, and at least more flexible, in relation to the many varied levels of interactivity, the many

Table 5.4: Bordewijk and Kaam's Matrix for the Four Communication Patterns

	Information produced by a central provider	Information produced by the consumer
Distribution controlled by a central provider	1) TRANSMISSION	4) REGISTRATION
Distribution controlled by the consumer	3) CONSULTATION	2) CONVERSATION

Source: Adapted from Bordewijk & Kaam, 1986

differing technologies, and rapid technological developments. It has also become clear that there are different forms of interactivity, which cannot readily be compared or covered by the same formula. There appears to be a particular difference in interactivity which consists of a choice from a selection of available information content: interactivity which consists of producing information via input to a system, and interactivity which consists of the system's ability to adapt to a user.

It might, therefore, be appropriate to operate with different—mutually independent—dimensions of the concept of interactivity.

The media typology developed by Bordewijk and Kaam[10] is based on two central aspects of all information traffic: the question of who owns and provides the information, and who controls its distribution. By cross tabulating these two aspects in relation to whether they are controlled by either a centralized information provider or a decentralized information consumer, a matrix appears with four principally different communication patterns, as illustrated in Table 5.4.

1) If information is produced and owned by a central information provider and this center also controls the distribution of information, we have a communication pattern of the **transmission** type. This is a case of one-way communication, where the significant consumer activity is pure reception. Examples would be classic broadcast media such as radio and traditional TV.

2) If information is produced and owned by the information consumers who also control distribution, we have a **conversation** communication pattern. This is a case of traditional two-way communication, where the significant consumer activity is the production of messages and delivery of input in a dialog structure. Typical examples would be the telephone, e-mail, and newsgroups.

3) If information is produced and owned by an information provider, but the consumer retains control over what information is distributed and when, it is a **consultation** communication pattern. In this case, the consumer makes a request to the information-providing center for specific information to be delivered. Here the characteristic consumer activity is one of active selection from available possibilities. Typical examples would be various on-demand services or online information resources such as FTP and websites.

4) Finally, if information is produced by the information consumer, but processed and controlled by the information-providing center, we have a **registration** communication pattern. In this communication pattern the center collects information from or about the user. In this case, the characteristic aspect is the media system's storage, processing, and use of the data from or about the user. Typical examples would be various types of surveillance, and logging of computer systems.

Among these four information patterns, transmission is the only one that is characterized by one-way communication. In other words, there is no return channel that makes information flow possible *from* the information consumer *to* the media system. Until now, communication and media studies have primarily based their models and insights on the transmission pattern because of the dominant role played by mass communication research. Current media developments including the arrival of "new media," such as the Internet, intranets, networked multi-media, and digital interactive television, have been more or less singularly characterized by a movement away from the transmission pattern toward the other three media patterns. These new media open up the possibility for various forms of input and information flow from users to the system: they can hardly be described using traditional one-way models and terminology. This is probably one of the main reasons for the lack of an appropriate and clear definition of interactivity in media sciences.

To go further, it seems appropriate to define interactivity as a continuum, where interactivity can be present in varying degrees. The degree to which a medium, or a mediated experience, can be said to be, and will likely be perceived as, interactive depends on (at least) five subsidiary variables.

1. The first variable is the **number of inputs from the user that the medium accepts and to which it responds**. There could in fact be a variety of user inputs, including voice/audio input (e.g., speech recognition systems that allow a computer to accept and respond to voice com-

mands), haptic input (e.g., television remote controls, knobs and buttons and computer mice, joysticks, wands, etc. that record user commands via object manipulation), body movement and orientation (kinetic) input (e.g., data gloves, body suits, and exoskeletons that translate body movements into electronic signals a computer can use to "fit" the user in a virtual environment), facial expressions and eye movements, and even psychophysiological input (e.g., heart rate, blood pressure, muscle tension, skin resistance, and brain waves could be input to a computer for mood management or enhanced mediated interpersonal communication).

2. **The number and type of characteristics of the experience that can be modified by the user** also help determine the degree to which a medium can be called interactive. Adjustable dimensions could be temporal ordering (order of events within a presentation), spatial organization (placement of objects), intensity (of volume, brightness, color, etc.), and frequency characteristics (timbre, color). Others might include size, duration, and pace. A greater number of the characteristics should generate perceptions of greater interactivity.

3. A third variable is the **range or amount of change possible in each characteristic of the mediated presentation or experience**. Expanding the degree to which users can control each attribute of the mediated experience enhances interactivity. For example, the larger the vocabulary of a computer speech recognition system (i.e., the more words it recognizes and to which it responds appropriately), the more interactive is the computer use experience.

4. A fourth variable is the **speed with which the medium responds to user inputs**. The ideal interactive medium responds in "real time" to user input; the response or lag time is not noticeable. Although it accepts and responds to only audio input and uses only a limited frequency range, the telephone is highly interactive in terms of this criterion because interactions via telephone seem to occur in real time (except with calls over exceptionally long distances). With bandwidth limitations and explosive growth in the number of users, the issue of response time is an important consideration on the World Wide Web.

5. A final variable that may be important for interactivity is the **degree of correspondence between the type of user input and the type of medium response**. Mapping between these two can vary from being arbitrary (e.g., pressing a sequence of keys on a keyboard to adjust a visual display) to natural (e.g., turning one's head in a virtual reality system to see the corresponding part of the environment). Using our familiar

sensorimotor skills to manipulate virtual objects directly may lead to perceptions of greater interactivity and naturalness than writing programs, twisting knobs, or pushing a mouse to accomplish the same task.

The concept of interactivity (as well as the concept of interaction) is complex and has a long list of very different, specific variations as the above review of the various concepts of interactivity has pointed out.

A definition as a continuum appears to be more appropriate, and at least more flexible, in relation to the many varied levels of interactivity, the many differing technologies and rapid technological developments. It has also become clear that there are different forms of interactivity, which cannot readily be compared or covered by the same formula. There appears to be a particular difference in interactivity that consists of a choice from a selection of available information content, interactivity that consists of producing information via input to a system, and interactivity that consists of the system's ability to adapt to a user. It might, therefore, be appropriate to operate with different—mutually independent—dimensions of the concept of interactivity. The various important aspects of the concept of interactivity can to a great extent be reduced to four dimensions which can be understood using the communication patterns: transmission, consultation, conversation, and registration.

Based on this understanding, interactivity may be defined as: *a measure of a media's potential ability to let a user exert an influence on the content and/or form of the mediated communication.*[11] This concept of interactivity can be divided up into four sub-concepts or dimensions, which could be called:

1) *Transmissional interactivity*—a measure of a media's potential ability to let the user choose from a continuous stream of information in a one-way media system without a return channel and therefore without a possibility for making requests (e.g., datacasting, multicasting, teletext, Near-Video-On-Demand).
2) *Consultational interactivity*—a measure of a media's potential ability to let the user choose, by request, from an existing selection of pre-produced information in a two-way media system with a return channel (e.g., WWW, Video-On-Demand, online information services).
3) *Conversational interactivity*—a measure of a media's potential ability to let the user produce and input his/her own information in the media system in a two-way media system, be it stored or in real time (e.g., videoconferencing systems, news groups, e-mail, mailing lists).

Figure 5.1: The "Cube of Interactivity": A Three-Dimensional Representation of the Dimensions of Interactivity

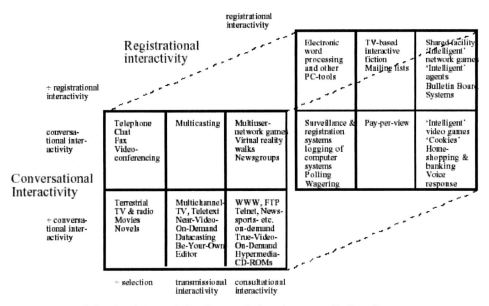

Source: Adapted from Intermedia White Paper on Interactive TV, 1999

4) *Registrational interactivity*—a measure of a media's potential ability to register information from and thereby also adapt and/or respond to a given user's needs and actions, whether they be the user's explicit choice of communication method or the system's built-in ability to automatically "sense" and adapt (e.g., surveillance systems, intelligent agents, intelligent guides, or intelligent interfaces). Since transmissional and consultational interactivity both concern the availability of choice—respectively with and without a request—it is possible to represent them within the same (selection) dimension.

The four types of interactivity can then be presented in a three-dimensional graphic model—an "interactivity cube"—as attempted in Figure 5.1, which in this form results in 12 types of interactive media.

CAPACITY: DOWNSTREAM AND UP-STREAM BIT-RATES

Whether networks are one or two way, the most useful method of categorising them is by the bit rate supported by the network—that is to say, the maximum data throughput expressed in terms of bits per second[12] (see Table 5.5).

Unlike the definitions of interactivity which are discontinuous, capacity can be expressed along a continuum. As a general guideline the term *high capacity* is used in this chapter to refer to any network supporting a bit-rate above

Table 5.5: Different Applications by Bit-Rate Required

Data speed	Quality	Application
2Mbit/s	VHS quality (analog video quality)	Channels with few moving pictures and/or little detail such as weather or cartoon channel
4-6 Mbit/s	Pal quality (analog TV quality)	Average TV channel with movement but no extreme requirements in relation to detail, such as news channel or film channels
8/9 Mbit/s	Pal quality	Sport channel with fast-moving action, such as an ice hockey match or skiing
	Studio quality	Average TV channel with movement but no extreme requirements in relation to detail, such as news channel or film channel
> 15 Mbit/s	High Definition TV (HDTV) quality	High Definition TV (HDTV) channel

Source: Adapted from Broadband Monitor, VECAI, Spring 2000

Figure 5.2: The Different Transfer Modes Capacities and Their Categorisation

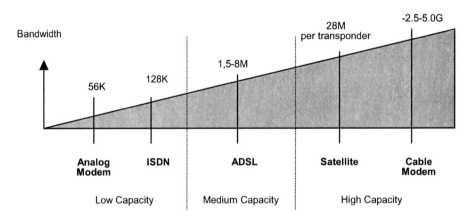

55Mbit/s, and *low capacity* to refer to any network supporting a bit-rate below 128 Kbit/s. Medium capacity means anything is between (Figure 5.2).

INTERACTIVE TELEVISION

The American Federal Communications Commission (FCC) defines iTV as "a service that supports subscriber-initiated choices or actions that are related to one or more video programming streams."[13] The British broadcasting regulator Independent Television Commission (ITC) defines interactive TV services as "pull" services initiated by the subscriber that are not necessarily related to any specific video programming. The ITC differentiates between two essentially different types of interactive TV services: dedicated and program-related services. The first are stand-alone services not related to any specific programming stream. Typically this will be entertainment, information, and transaction services. Program-related services refer to interactive TV services that are directly related to one or more video programming streams. These services allow users to obtain additional data related to the content (either programming or advertising), to select options from a menu, to play or bet along with a show or sports event, to interact with other viewers of the same program.

On the basis of the terminology adopted in this chapter, interactive television is defined as domestic television boosted by interactive functions, made possible by the significant effects of digital technology on television transmission systems.[14]

A first level of analysis shows that interactive television is a system through which the viewer can ask something to the program provider. In this way the viewer can transmit his own requests through the two-way information flow, made possible by the digitalisation of the television signal.

The viewer's reception of the digital signal is made possible through a digital adapter (*set top box or decoder*) which is connected to the normal television set or integrated with the digital television in the latest versions. The set top box decodes the digital signals in order to make them readable by the conventional analog television set (Figure 5.3). The set top box has a memory and decoding capacity that allows it to handle and visualise information. Thus, the viewer can accede to a simple form of interactivity by connecting the device to the domestic telephone line. In addition, other installation and infrastructure arrangements are required, depending on the particular technology. In particular a return channel must be activated. This can imply a second dedicated

Figure 5.3: Interactive Television: How it Works

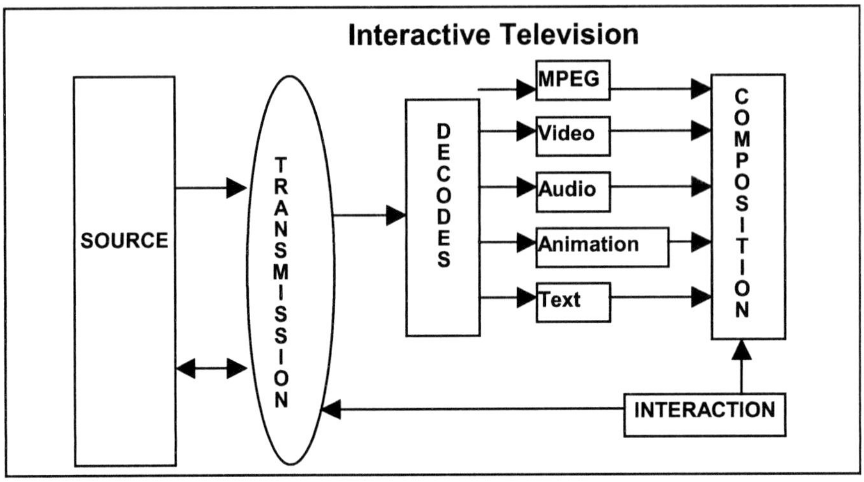

```
┌─────────────────────────────────────────────────────────────┐
│                    Interactive Television                    │
│                                    ┌─MPEG──┐                 │
│  ┌────────┐      ┌───┐    ┌───┐   ├─Video─┤    ┌───┐        │
│  │        │      │ T │    │ D │   ├─Audio─┤    │ C │        │
│  │ SOURCE │ ───▶ │ R │───▶│ E │──▶├─Anim──┤───▶│ O │        │
│  │        │      │ A │    │ C │   └─Text──┘    │ M │        │
│  └────────┘  ◀──▶│ N │◀───│ O │                │ P │        │
│                  │ S │    │ D │                │ O │        │
│                  │ M │    │ E │   ┌──────────┐ │ S │        │
│                  │ I │    │ S │   │INTERACTION│─│ I │        │
│                  │ S │◀───┴───┴───│          │ │ T │        │
│                  │ S │            └──────────┘ │ I │        │
│                  │ I │                         │ O │        │
│                  │ O │                         │ N │        │
│                  │ N │                         └───┘        │
│                  └───┘                                       │
└─────────────────────────────────────────────────────────────┘
```

Source: Adapted from CSELT, 1998

telephone line for return path via modem. The end user can interact with his TV set through a special remote control, or in some cases even with a wireless keyboard.

This means that, even in houses where there is no personal computer, online services, such as the Internet, can still be accessed. Through the television, interactive services can be used separately from, or combined with, television transmissions, for example to enhance television programs through the provision of information on request. In this way the viewer can interact with the contents of television programs, using the telephone to send response messages to the service provider or to advertisers. This process facilitates services such as home shopping or electronic banking and others.

NEW OFFER TYPES

Interactivity is a functionality rather than a specific type of service, and it can be applied in a wide variety of contexts. Its distinguishing characteristic is the ability of viewers to interact with TV programs by one of two methods:

- by changing the content which appears on the screen—for example to access background information, to change camera angles, to view more

than one picture at a time, or to view associated text at the same time as a main picture;
* by providing information to the broadcaster through a return path, usually a telephone line—for example to order a product, to exercise "votes" on options provided by a program or to participate in an on-screen quiz show.

These services are available only to members of the public with digital equipment, whether satellite, cable, or digital terrestrial. A range of interactive services is now being provided to the public by broadcasters using both of these types of interactivity, although many projects are still at the pilot stage.

A separate but related development has been the growth of Internet-via-TV services. These provide access to the full World Wide Web rather than to content moderated by the broadcaster—such Web access is not normally edited, except to the extent necessary to make text legible on a TV screen. This

Figure 5.4: The Evolution of Interactive Digital Television

Analogue Broadcast With embedded digital data	Digital Broadcast With data content service	Digital Broadcast & Back channel	Digital Broadcast, Back channel, Local storage	Digital Broadcast Broadband Backchannel Local storage
		EVOLUTION OF INTERACTIVE DIGITAL TELEVISION		
Subtitles on/off E.g. Ceefax Analogue Satellite conditional access Pay to view channels	Enhanced information services; STB Interactivity; EPG's, PPV movies, Games	Back channel interactivity, Impulse PPV movies, Impulse shopping, On line multiplayer gaming, voting, RELIANCE on telephone connections	Non-linear broadcasting, personal video recorder, skip advert, Limited video on demand, personalised content, targeted advertising, decreased reliance on phone back channel	True video on demand always broadband connection dedicated back channel, video conferencing, home network, interactive live TV, true audience participation

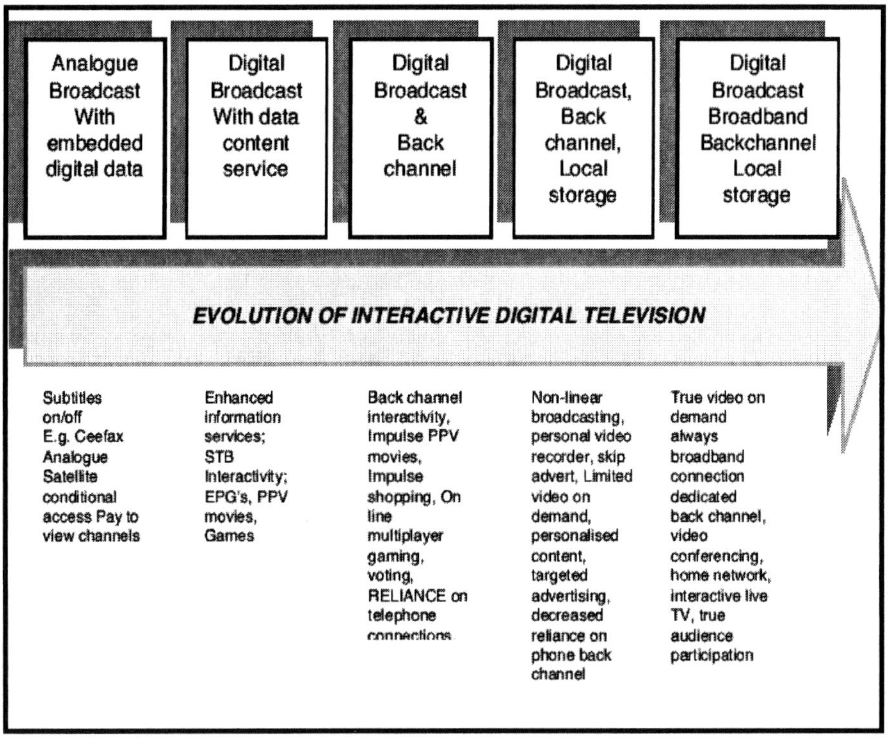

Source: Adapted from Arthur Andersen

kind of service uses the TV essentially as a computer screen, in connection with a set top box providing a function similar to a PC.

How does the traditional broadcaster's offer change through the use of new interactive technologies?

Content offered to the user by interactive television keeps, as in the case of the traditional analog television, the intrinsic feature of collectively consumed service[15] (although non-payers can be cut off from it); yet this feature is lost when contents are broadcast. Each interactive television user can indeed request a separate and customised program, and once broadcast, the content has features much like a private good and/or service.[16]

Consumers have access to a range of interactive services which can be classified into some main categories (Table 5.6) characterized according to different interactive applications such as enhanced TV, Pay-Per-View (PPV), interactive game shows, communication, TV banking and interactive finance, home shopping, interactive advertising, and Web access.

The results of an analysis[17] conducted in Western Europe in 2002, on a sample of 607 interactive services offered by digital television platforms, show that 40% of the total interactive services are enhanced TV services (Figure 5.5).

Table 5.6: Interactive Television Services

Category	Interactive application
Enhanced TV	Personalised weather information
	Personalised EPG (Electronic Program Guide)
	Menu à la carte
	Different viewing angles
	Parental Control
	Enhanced TV
	Multi language choice
Games	Single-player games
	Multi-player games
	Voting and betting
Communication	Instant messages
	E-mail
Finance	Financial information
	TV banking
E-commerce	Pay-Per-View
	Home shopping
Advertising	Interactive advertising
Internet	Web access

Figure 5.5: Interactive Television Services Offered in Europe 2002

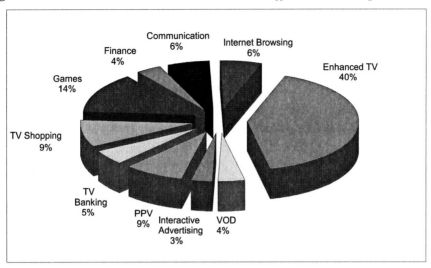

Source: New Media&TV-Lab, I-LAB Bocconi University, 2002

These services can be characterized according to their relation to TV programs and their relation to TV commerce.

TV commerce is here considered in its larger sense, in relation to the purchasing cycle which includes advertising, promotion, purchasing, and after-sale services.

Most of the TV commerce services are business-to-consumer oriented, although signs of business-to-business services have also been found.

Three kinds of television have often been contrasted: television of culture, television of entertainment, and television of information. TV programs can be situated in this triad: Information, Play, and Fiction. Due to the pervasiveness of commerce in interactive television, Commerce has been added to qualify interactive services (see Figure 5.6)

Different interactive applications have different profitability levels (Table 5.7); Video-On-Demand, Pay-Per-View, home shopping, TV banking, and interactive advertising offer the best potential earnings, while different forms of enhanced TV appear to be less so.

The major critical success factor for all the categories of services considered is their content, which makes them attractive face-to-face TV viewers. New ideas and the production of new contents are a must to have a larger audience, and above all TV viewers must be aware of what is being offered to them.

Figure 5.6: Interactive Television

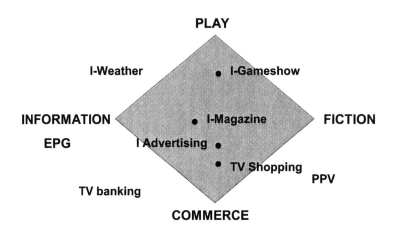

Table 5.7: Earning Potential for Interactive Applications

% of answers included in the ranking of high potential revenue*	
Video-On-Demand	75.0%
E-Commerce	61.7%
Interactive Advertising	53.3%
On-Demand Programs	41.7%
News & Information	38.3%
Interactive Games	36.7%
Personal Video Recorders (PVRs)	33.3%
Interactive TV Guide	30.0%
Web Video Streaming	23.3%
Enhanced TV	20.0%
Interactive Applications in TV Shows	20.0%

*Ranking of 6 or 7 over 7 points
Source: Adapted from Meyers Group Prospects Data for ITV, 2000

Interactive services are gradually becoming an essential feature of digital TV services and will attract a greater number of subscribers and create a new source of income and competitive advantage.

ELECTRONIC PROGRAM GUIDE AND PAY-PER-VIEW

Electronic Program Guide (EPG) and PPV applications are considered separately from interactive programming because, although they do belong to the television sphere, they are not attached to specific TV programs but qualify as meta-program. EPG relates to the description of TV programs and acts as a TV navigation system or portal. PPV can be viewed as an alternative model to subscription.

Electronic Program Guide

EPG or electronic program guide is the channel selection device at the heart of the digital TV revolution. EPG is an essential, navigational device allowing the viewer to search for a particular program by theme or other category and order it to be displayed on demand. Ultimately, EPGs will enable the TV set to learn the viewing habits of its user and suggest viewing schedules.

EPG (or IPG—interactive program guide) helps people grasp a planning concept, understand complex programs and absorb a large amount of information quickly, and navigate in the TV environment. In the future the EPG may become the preferred entrance to different media such as TV, radio, websites, and interactive services.

Most EPGs are built in the same way: viewers can browse by channels, program categories, or titles. Typical features are flip, browse, and video browser. The flip bar appears every time the channel is changed, displaying the current channel, the name of the program, and its start and end time. The browse feature allows viewers to see program listings for other channels without missing a minute of the TV program (Startsight). In an interactive video browser, viewers see thumbnails of TV programs when entering a specific description the title and sound of the current program is given (MediaHighway). Another navigation method includes search by keywords; other features are multi-language choice and VCR programming.

More advanced features under development concern customization and personalization. Customization starts with features like favorites, or reminders, which can be set for any future program. The reminder appears a few minutes before the program starts, allowing the viewer to turn to that program. Viewers can designate channels as Favorites and subsequently quickly navigate to those channels.

An extension of this idea is personal profile. This is a very interesting area for future development, for which there are two main development directions: ranking systems and noise filter. Ranking systems are seen as preference systems, where viewers can order channels, from the most watched to the least watched. Noise filters are seen as systems in which viewers block information, for example removing channels that they never watch.

One related issue is parental control (filter), where objectionable programming can be restricted by setting locks on channels, movies, or specific programs.

To choose a program, viewers can further consult descriptive information including synopsis, actors, genre, audience acceptance, and duration. Thus another direction for development is the provision of in-depth program information, or interactive TV magazine. Similarly to printed TV guides, it could contain qualitative information such as ranking lists, tips of the day, reviews, and preview clips.

The EPG can be seen as the lobby for viewers to move through different channels or services.

Pay-Per-View

Consumers can also select Pay-Per-View (PPV) services which are offered by a number of broadcasters (TPS, BSkyB, BBC, etc.). Similarly to computer software download, viewers can select Video-On-Demand (VOD).

Video-On-Demand technology provides an alternative to the broadcast environment, and through broadband[18] connections offers viewers on-demand access to a variety of server-based content on a nonlinear basis.

Viewers can choose to pay for specific live-sport events or spectacles such as theatre or dance. Most companies have developed specific promotional features or channels designed to attract customers. Such features can include video clips, text billboards, and can be supported by discount tokens or season tickets (e.g., for football matches and Formula 1 racing).

INTERACTIVE TV SERVICES

As described above there are two essentially different types of interactive TV services: dedicated and program-related services.

Dedicated services are stand-alone services not related to any specific programming stream. They follow a model closer to the Web even if there are differences in: hyperlinks, media usage, and subsequently, mode of persuasion. This type of interactive service includes entertainment, information, and transaction services.

Program-related services refer to interactive TV services that are directly related to one or more video programming streams. These services allow users to obtain additional data related to the content (either programming or advertising), to select options from a menu, to play or bet along with a show or sports event, to interact with other viewers of the same program.

Interactive Games

Interactive games, similar to small Web games integrated into an arcade game, are also proposed as part of the interactive channels.

Interactive game shows take place in relation to game shows, to allow viewers to participate in the game.

A new emerging type of interactive games are network games, which allow teleplayers to compare scores and correspond by a form of electronic mail, or to compete against other players.

There are different revenue models related to the offer of games: subscription fee, pay-per-play or pay-per-day, advertising, sponsorship, and banner, as shown in Figure 5.7.

Figure 5.7: Revenue Model Related to the Offer of Interactive Games in Europe (Base 85 Interactive Games Analysed)

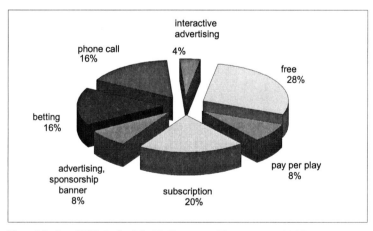

Source: New Media &TV-Lab, I-LAB Bocconi University, 2002

Interactive Advertising

Interactive advertising is synchronized with the TV ad. An interactive overlay or icon is generated on the screen leading to the interactive component. When the specific pages are accessed, viewers can learn more about products, but generally other forms of interactions are also proposed. Viewers can order catalogs, can benefit from a product test, and can participate in competition, draw, or play games. It could be interesting to explore other possibilities, especially with regard to advertised products and viewers. The interactive ad should be short in order not to interfere with the program that viewers wish to watch. The message must be simple and quick. This strategy is based on provoking an impulsive response (look at the interactive ad) resulting in the required action (ordering the catalog). A natural extension of this concept is to enable consumers to order directly.

Originally it was thought that high-info and expensive products like cars, banking, hi-fi requiring a complex commercial argumentation should prove the best candidate for this kind of advertising. However, it has been shown that this kind of advertising would work with all kind of products. The results of an analysis[19] conducted in Europe in 2002, on a sample of 136 interactive advertising services offered by digital television platforms, show that grocery (19%), cars and accessories (15%), banking and financial services (13%), audiovisual products and entertainment (11%) are the classes of products which more frequently use interactive television to promote their products and services (Figure 5.8).

Figure 5.8: Interactive Advertising: Kind of Products (Europe 2002)

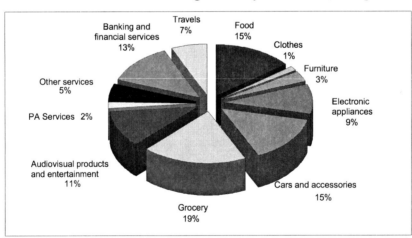

Source: New Media&TV-Lab, I-LAB Bocconi University, 2002

One question under discussion is the impact of this kind of advertising on other TV ads or programs. Interactivity might demand that ad messages be reformulated. This is essential for talking differently of a product, to create a complicity with consumers so that they can appropriate the brand. The communication must adapt and reinvent itself; it is necessary to create another communication technique. Integrating TV and multimedia will lead to a marketing of mass customization.

TV Shopping

TV shopping is common both on regular channels and on specialized channels. Some channels specialized in teleshopping (QVC in USA and Europe, and Home Shopping Europe in Europe) have also a Web presence. We are at the very early stage of such systems; French consumers have been able to buy (via TPS boutique) toys (Lego), foods (CASINO), and CDs and books (FNAC), all well-known brands.

TF1, a French channel via TPS, has developed an interactive teleshopping program. Consumers can order products currently shown in the teleshopping program and pay by inserting their credit card in the set top box card reader. During the program, an icon appears signaling to viewers that they can now buy the item.

The chosen product is then automatically displayed in the shopping basket. Viewers enter the quantity and the credit card number. The objective of such a program is to give viewers the feeling of trying products. The products' merits are demonstrated in every dimension allowed by the medium. In some ways we can consider teleshopping as the multimedia counterpart missing from Web shops. Instead of being faced with endless rows of products displayed in the same format, consumers can now see the product in use, develop a feeling for the product and being tempted to buy the product.

We should not minimize the value of this kind of program as entertainment. The program takes the form of a talk show, usually presented by famous hosts interacting with an audience (guests, studio audience, and outside callers).

Mixing elements of teleshopping and e-commerce might constitute a useful example of integration of TV and interactivity, resulting in a new form of interactive shopping. Consumers can be enticed by attractive features and seductive plots.

There is a difference between interactive advertising and interactive shopping. Initially, interactive advertising is triggered from an ad and concerns a specific product. Shops on the other hand are accessed directly from the TV shopping section and concern a range of products.

TV shopping presents a business model close to PPV and has a huge potential.

TV Banking

TV banking services are becoming a prominent feature of the iTV landscape, with providers in different countries intending to set up such a service. The UK, Spain, and France are the countries with the largest offering of TV banking interactive services[20] (Figure 5.9).

TV banking enables consumers to consult their bank statements, and carry out their day-to-day banking operations. However, other financial services are also provided in this framework—consumers can carry out financial operations, receive personalized investment advice, or consult the Stock Exchange online.

Interactive TV gives financial services companies new scope for marketing: it permits them to display their products in full-length programs rather than commercials lasting a few seconds, and to deliver financial advice in interactive formats, even in real time. Such companies particularly value the ability to hot-link traditional TV commercials to sites where viewers can buy products online. In addition, service providers on interactive TV can tailor their offers precisely by collecting detailed data about the way customers use the medium.

Figure 5.9: TV Banking Interactive Services Offered by Digital Platforms in Europe 2002

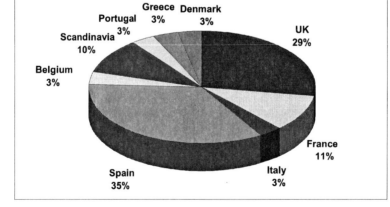

Source: New Media&TV-Lab, I-LAB Bocconi University, 2002

Even if interactive TV reaches technical parity with the PC as a Web access device, financial services providers will have to design custom-made propositions to compete, for their current, text-heavy websites and product-based offers look dowdy in an entertainment medium. Designing online services for TV requires video and content development skills that few banks have in-house, requiring them, in all likelihood, to join forces with television and media specialists.

Model of Interactive Programming

Interactive programming works by:

- adding layers of depth to what is happening on the screen (the most common model);
- adding details or content that could not fit within the show's original running time, production schedule, or format (a model for DVD, although no evidence has been found regarding iTV);
- rewarding the audience for good behavior (promotion, and perhaps games fit this category);
- supporting TV commerce (the more pervasive form).

Interactive programming should enhance the TV viewing experience. It often occurs as a natural extension of the TV genre, enhancing the primary characteristics of the program, thus facilitating viewers' assimilation of this new form. By allowing each viewer to participate in the game, the experiential quality of the game show program (play) is reinforced, thus boosting the audience interest and engagement. Similarly, buying or ordering a catalog after an advertisement seems a natural extension of the form and function of this kind of program (commerce).

Due to their audience (being high spenders) and success with other electronic medium (the Web), some specialized channels such as MTV seem prime candidates for interactive programming. Numerous documentaries and news programs could benefit from the addition of an informative interactive module, by clarifying the program content with the aid of maps, diagrams, charts, or any kind of encyclopedia information.

The best way to find out what kind of interactivity could be added to a specific program (especially in relation to TV commerce) is to look at consumer behavior and patterns of consumption: what forms of media are consumed in a specific area, what kinds of goods are purchased in relation to the domain, what kind of activities do people enjoy doing and in what context? What is acceptable or enjoyable will be determined by the viewer's motivation.

At the moment, only very few possibilities of interactive programming have been explored. Most interactive programs occur concurrently with TV programs, for example viewers play at the same time as the game show, viewers can potentially buy while the teleshopping show continues in the background.

By contrast, interactive advertisements imply a disruption of the program, however brief. The viewers cannot watch the program and interact with the ad at the same time. Viewers are leaving the TV space to enter interactive ad space (DAL—Dedicated Advertiser Location), however they do remain in the interactive programming space, a sharp distinction with interactive services and the Web. We can envisage different and more complex forms of interactive programming where the genre of the interactive and TV program are different (e.g., TV commerce) or that one TV program would have multiple genres of interactive programs.

By looking at TV commerce, it is possible to imagine all kinds of scenarios; it is always possible to sell something related to a TV program: to sell newspapers with news programs, TV goods with series, and the interactive weather program could be linked to specific news programs and be sponsored.

A variation on TV commerce has been found in relation to multicasting. Multicast broadcasting allows a new way of viewing. Instead of a single linear stream of information, a digital TV program can consist of concurrent multiple audio-video stream broadcasts. Multiple camera angles or other data channels can be displayed. If one channel is reserved for TV commerce, the performers' clothes would be offered for sale. All programs could be linked to a TV commerce channel.

If we consider the idea of a TV portal, all TV programs could be embedded in a framework including four functions: PROG, PLAY, INFO, BUY.

- Program (PROG) would cover smart features associated with the EPG.
- PLAY would include any function related to participation and conversation (game, vote, chat).
- INFO would include acquisition of information related to the program content.
- BUY would cover TV commerce.

Interactive programs could be generated by addition, substitution, repetition, or transformation to the TV program. New interactive programming will demand a deep knowledge of the television world as well as games, e-commerce, consumer behavior, and audience behavior.

INTERACTIVE DIGITAL TELEVISION (IDTV) VALUE CHAIN

The interactive digital television marketplace is complex, with competing platforms and technologies providing different capabilities and opportunities. Interactive television has a larger number of key stakeholders and a more complicated set of processes and relationships than traditional TV.

The multi-channel revolution, coupled with the developments of interactive technology, is truly going to have a profound effect on the supply chain of the TV industry. Interactivity only partially shows the strong innovation in the content industry, in both production and services as well as in operating activities and management styles. Digital technologies have a large effect at every stage of the value chain for television broadcasters (from production of programs to their distribution).

The competitive development generated by interactivity also creates new business areas, requiring new positioning along the value chain[21] for existing operators.

Table 5.8: The iDTV Value Chain: Players and Added Value

Player	Added Value
Content provider	Produce content and edit/format content for different iDTV platform
Application developer	Research and develop interactive applications
Content aggregator	Acquire content rights, reformat, package, and re-brand content
Network operator	Maintain and operate network, provide adequate bandwidth
iDTV platform operator	- Acquire aggregated content and integrate into iDTV service applications - Host content/outsource hosting - Negotiate commerce deals - Bundle content/service into customer packages - Track customer usage and personalise offering
Customer equipment	- Research and development equipment - Manufacture equipment - Negotiate deals and partnerships

Several types of companies are involved in the iDTV business: content provider, application developer, television producers and broadcasters, network operator, iDTV platform operator, hardware and software developers, Internet developers also interested in developing for television, consultants, research companies, advertising agencies, etc. (Table 5.8).

A central role is played by broadcasters whose goal is to acquire contents from content providers (banks, holders of movie rights, retailers), to store them (storage), and to define a broadcast planning system (planning). They directly control users' access as well as the quality of the service and its future development (see Figure 5.10).

Conditional access is an encryption/decryption management method (security system) through which the broadcaster controls the subscriber's access to digital and iTV services, such that only those authorised can receive the transmission. In addition to encryption/decryption of the channel, conditional access services currently offered include security in purchase and other transactions, smart card enabling and issuing, and customer management services (billing and telephone servicing). The subscriber most often uses "smart cards" and a private PIN number to access the iTV services. Not all services are necessarily purchased from the conditional access operator. Some conditional access systems are: NDS' Open VideoGuard, Canal+ Technologies' Mediaguard, Irdeto, SkyStream's DVB-Simulcrypt and Microsoft's Access. Mediaguard and Viaccess appear to be widely used throughout the European Union.

Service providers such as data managers provide technologies that allow the broadcaster to deliver personalized, targeted content. They use a Subscriber Management System (SMS) to organize and operate the company business. The SMS contains all customer relevant information and is responsible for keeping track of placed orders, credit limits, invoicing and payments, as well as the generation of reports and statistics. Industry leaders are DoubleClick, Jupiter Media Matrix, Nielsen Netratings, and Predictive Networks.

Satellite platforms, cable networks, and telecommunications operators, mainly focused on the distribution of the TV signal, gradually tend to integrate upstream in order to have a direct control over the production of interactive services.

The vast **end device** segment includes two sub-segments regarding the hardware and the software embedded in it.

The hardware manufacturers (such as Sony, Philips, Nokia, and others) design, produce, and assemble the set top boxes (STBs). The set top box is the

Figure 5.10: The iDTV Value Chain: Head-End Phases

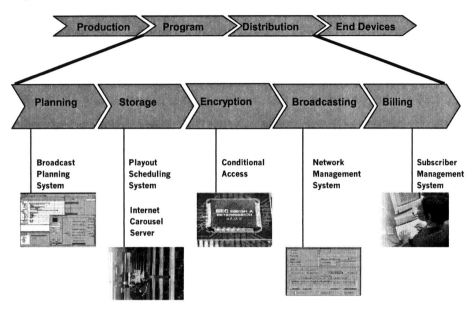

physical box connected to the TV set and the modem/cable return path. It decodes the incoming digital signal, verifies access rights and security levels, displays cinema-quality pictures on the TV set, outputs digital surround sound, and processes and renders the interactive TV services.

The *software* sub-segment includes:

1. *Operating systems* (OSs) developers provide many services, such as resource allocation, scheduling, input/output control, and data management. Although operating systems are predominantly software, partial or complete hardware implementations may be made in the form of firmware.[22] The main OSs developed are Java Virtual Machine by SunMicrosystem, Windows CE by Microsoft, and Linux.
2. *Middleware* providers and developers provide programming that serves to "glue together," or mediate, between two separate and usually already-existing programs. Middleware in iTV is also referred to as the Application Programming Interface (API): it functions as a transition/conversion layer of network architecture that ensures compatibility between the basal infrastructure (the Operating System) and diverse upper-level applications. There are four competing technologies: Canal+ Media Highway (running on Java OS), Liberate Technologies (Java), Microsoft TV

Figure 5.11: The iDTV Value Chain: End Device

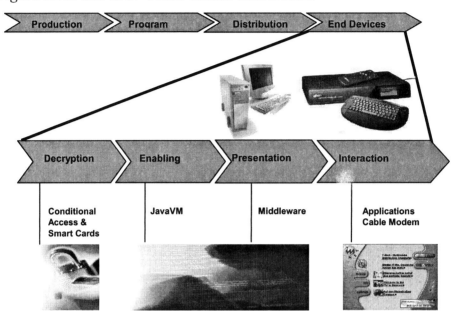

(Windows CE), and OpenTV (Spyglass). These are all proprietary solutions acting as technological barriers and trying to lock-in the customers. This situation creates a vertical market where there is no interoperability, and only programs and applications written specifically for a system can run on it.

3. The *user-level applications provider* category includes interactive gaming, Interactive (or Electronic) Programming Guides (IPGs or EPGs), Internet tools (email, surfing, chat, instant messaging), t-commerce, Video-On-Demand (VOD), and Personal Video Recording (PVR).

BECOMING A CUSTOMER OR CONTENT GATEWAY IN THE NEW ECONOMY

With reference to the flow of revenues among the different operators in the different stages of the value chain (Figure 5.12), the network operators have an advantage in gaining revenue because interactive-TV applications such as enhanced broadcasting are tightly linked to the network infrastructure. Their condition of "bottleneck," through which all the content that appears on

Figure 5.12: Revenue Flow Among the Value Chain Elements

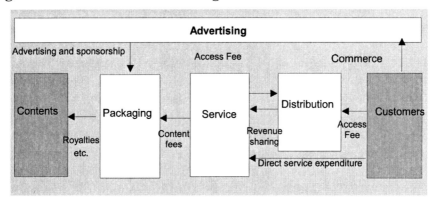

Source: Adapted from Arthur Andersen "Competing in the Digital Economy," 2000

television must pass, provides a steady revenue stream for Pay TV services while interactive TV applications are being developed.

Companies operating in the "service" element collect their revenues directly from final users through their services. Revenues are then distributed among the other upstream stages of the value chain. Organisations active in the packaging stage (broadcasters) gather relevant additional revenues from indirect sources, such as advertising which represents a fundamental part of the revenue model.

Within telecoms, the percentage of call revenues funded by advertisers (through freephone, national call rate, and local rate numbers) continue to rise.

Some critical success factors, for the purpose of achieving and holding on to long-lasting competitive positions, are: the gradual control of the contents, thus becoming *content gatekeeper,* as well as by the strengthening of the control over final users' access, as *customer gatekeeper* (Figure 5.13).

Acquiring a position of *content* and *customer gatekeeper* gives the user a better chance to determine the ways through which to develop, assemble, sell, and distribute both the content and services.

Yet, it is worth pointing out that contents do indeed represent the actual killer application in the new digital TV market and that hypercompetition exists among channels to gain access to contents which most appeal to viewers.

In some cases, acquiring control positions for the access of the final users (as a customer gatekeeper) was coupled by a parallel upstream integration of content control leading to the integration of different stages of the value chain of the same entity (e.g., BskyB in the UK straddles significant portions of the value chain).

Figure 5.13: The Role of Content and Customer Gatekeeper

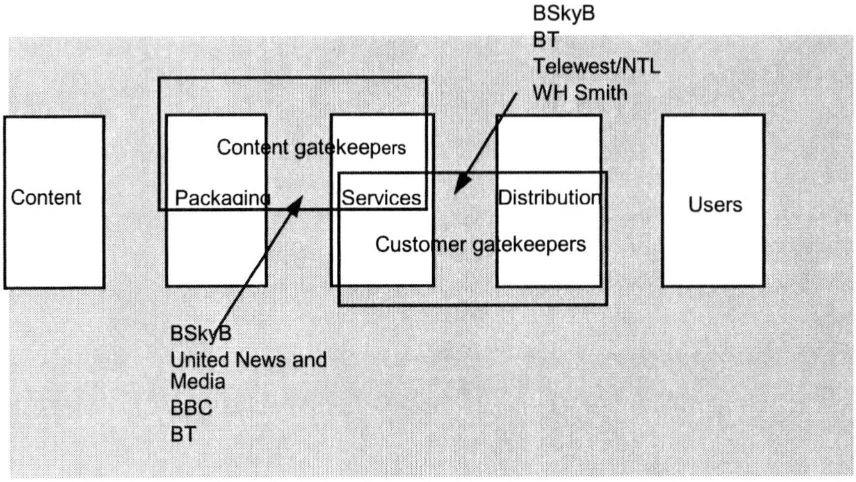

```
                                      BSkyB
                                      BT
                                      Telewest/NTL
                                      WH Smith

           Content gatekeepers
 ┌────────┬──────────┬──────────┬──────────┬────────┐
 │Content │Packaging │ Services │Distribution│ Users │
 └────────┴──────────┴──────────┴──────────┴────────┘
                       Customer gatekeepers

          BSkyB
          United News and
          Media
          BBC
          BT
```

Source: Adapted from Arthur Andresen "Competing in the Digital Economy," 2000

Digital television will accelerate television into a more convergent environment. The Digital Television market will be fundamentally more open than its analog forerunner, with intensified competition on the demand side for consumer ownership and on the supply side for content ownership. New business models and forms of commerce will evolve rapidly.

Today's television leaders face a dilemma about the roles they want to play in the digital television landscape of the next five to ten years. Essentially, this boils down to a simple question: do they want to be primarily a world class *content* owner, or world class *consumer* owner?

The first option involves developing and exploiting branded content assets (as discussed in Chapter 6) through a range of consumer distribution channels and formats—an approach often summed up as "content is king." Today an example of this approach is Discovery.

The second option involves building a business around television-based consumer relationships—so the consumer is king.

Examples of "consumer owners" today include most European cable companies, which look to capitalise on control of the "golden mile" and the ultimate direct link with the consumer.

Digital Television players will often need to decide whether to focus primarily on consumer ownership or content ownership. It will be very difficult to be world class at both. Perhaps only a few can have the scale. Mediocrity in each is likely to be squeezed out. By implication, non-world class or subscale television companies with activities strung along the content-to-consumer value chain face change or decay.

However, some degree of vertical integration will be a key supporting strategy for both "content owners" and "consumer owners." Consumer owners can use their own content to build a unique offer, and content owners can use direct consumer relationships to keep in touch with their markets. This is vertical integration that supports a focused consumer or content strategy.

ENDNOTES

[1] The word "use" is appropriate for interactive television, presuming there will be at least some changes in our television behaviour that extend our traditional television watching with activities which require a more active behaviour.

[2] From the terminological viewpoint the definition of multimedia service adopted refers to a type of service which includes more than one type of information (text, audio, pictures and video), transmitted through the same mechanism and allowing the user to interact or modify the information provided.

[3] Durlak, J., "A typology for interactive media", in McLaughlin, M. (ed.), Communication Yearbook 10, Newbury Park, Sage publications (1987).

[4] Carey, J. (1989) "Interactive media, International encyclopedia of communications", New York, N.Y., Oxford University Press.

[5] Rogers, E. (1986) *Communication technology. The new media in society*, New York, N.Y.

[6] Szuprowicz, B. (1995). *Multimedia Networking.* New York: McGraw-Hill.

[7] Laurel, B. (1991). *Computers as Theatre.* Reading, MA: Addison-Wesley.

[8] Goertz, L. (1995). *Wie interaktiv sind Medien?* In Rundfunk & Fernsehen, No. 4 (1995).

[9] Heeter, C. (1989). Implications of new interactive technologies for conceptualizing communication. In Salvaggio, J. & Bryant, J. (Eds.), *Media Use in the Information Age: Emerging Patterns of Adoption and Consumer Use.* Hillsdale, NJ: Lawrence Erlbaum Associates.

10 Bordewijk, J. & Kaam, B. (1986). Towards a new classification of teleinformation services. Inter Media, 14,1, as cited in White Paper on Interactive TV.

11 Intermedia. (1999). *White Paper on Interactive TV.*

12 To give some idea of the variation, a phone line typically supports a maximum rate of 64 Kilobits per second (64.000 bits per second) in either direction. A typical satellite transponder, by contrast, might support 28 Megabits per second (28.000.000 bits per second) down to the dish.

13 FCC. (2001). *In the Matter of Non-Discrimination in the Distribution of Interactive Television Service over Cable.* CS Docket, No. 01-7, p.2.

14 The impact of digital technology is the first determinant for the evaluation of the technological asset as the cause of structural and competitive change in the television market. In this regard the technological change generated by the advent of the new digital technologies shows, according to the approach used, the important economic components. Vaccà shows the methodological need to investigate the characteristics of the new technologies and their economic and competitive implications. Vaccà, S. *Imprese e Sistema Industriale in una Fase di Rapida Trasformazione Tecnologica...*, op. cit., pp. 78-79.

15 In the case of collectively consumed services, the consumption by the single user is compatible, or better not rival, with the consumption by one or more individuals. Brusio, G. (1993). *Economia e Finanza Pubblica,* Rome, Italy: NIS La Nuova Italia Scientifica, p. 55.

16 Owen, B.M. (1999). *The Internet Challenge to Television:* Cambridge, MA: Harvard University Press, pp. 63-64.

17 Source: New Media&TV Lab, I-LAB Research Center on Digital Economy, Bocconi University, July 2002.

18 Broadband—A network capable of delivering high bandwidth. Broadband networks are used by Internet and cable television providers. For cable, they range from 550 MHz to 1GHz. A single TV regularly broadcasting a channel requires 6MHz, for example. In the Internet domain, bandwidth is measured in bits-per-second (BPS). See DSL.

19 Source: New Media&TV Lab, I-LAB Research Center on Digital Economy, Bocconi University, July 2002.

20 Source: New Media&TV Lab, I-LAB Research Center on Digital Economy, Bocconi University, 2002.

21 "A firm's co-ordinated set of activities to satisfy customer needs, start[s] with [a] relationship with suppliers and procurement, going through production, selling

and marketing, and delivering to the customer. Each stage of the value chain is linked with the next stage, and looks forward to the customer's needs, and backwards from the customer too. Each link of the value chain must seek competitive advantage: it must be either a lower cost than the corresponding link in competing firms, or add more value by superior quality or differentiated features. The basic idea behind the value chain was first made explicit by Michael Porter in 1980" (Koch, 2000).

[22] Firmware is a program that is permanently stored in the electronic circuit boards of a computer, or in the Read-Only Memory and cannot be changed by a user.

PART III

MANAGING THE OPPORTUNITIES CREATED BY DIGITAL CONVERGENCE

Chapter VI

Branding Strategies for Digital Television Channels

NEW SOURCES OF COMPETITIVE ADVANTAGE BECOME ESTABLISHED

In the previous chapters, technical features and economic implications following the digitalisation of the TV signal as well as the development of interactive television were the focus of our analysis. At this point, it is worth considering some managerial implications stemming from the adoption of these new digital technologies.

The goal of this section is to determine those managerial areas mainly influenced by the digitalisation process, as well as the way corporate strategies define changes.

In this chapter, focus is placed on the analysis of the impact of digitalisation on marketing strategies through an investigation on the growing importance of the brand as a loyalty-based resource, available to digital television networks to aggregate and make loyalty vis-à-vis television viewers more concrete.

Special attention is being paid to branding policies adopted by digital television networks through a better knowledge of the reasons why brand equity is important in the television industry.

The trend towards a progressive worsening of the competition pressure in the television industry results from a number of interconnected causal factors which can generally be reconciled based upon the huge technological and competitive changes in this industry.

As it was pointed out in the previous chapters, the advent of the new digital technologies and the convergence process within *Information Communication Technology* (ICT) all caused progressive hyper-competition which forced each single broadcaster as well as digital platforms to create strong brand identities.

Broadcasters are increasingly gaining control over the "personality" of their networks and on viewers' perception of underlying "brand values." Therefore, the critical aspects of this analysis make one understand the origins of this hyper-competition process and identify the tools a network may need to pursue a successful brand strategy allowing for market competition and competitive position defense over time.

In this phase of the study, following a quick analysis on market evolution in terms of types and existing networks, the cognitive process adopted by a viewer when selecting a channel is considered, as well as the branding strategy and the tools that a television network and an iTV portal adopt to communicate values connected with their brand.

The goal of this analysis is to try to understand how a digital television network may create a channel experience through increasing viewers' loyalty as the source of the competitive advantage.

THE CUSTOMER-BASED BRAND EQUITY CONCEPT

According to a customer-based perspective,[1] brand equity may be defined as the differential impact that brand knowledge can produce on customers' response to a product and marketing policies by the brand itself (Busacca, 2000).

Brand knowledge components of brand equity may be identified as follows:

- brand identity;
- brand awareness;
- brand image.

Brand identity includes entrepreneurial values which are the basis for the existence of the brand, as well as all the elements to ease its recognition and memory (name, logo, symbols, jingles, slogans, etc.).

Brand awareness refers to the strength of brand knowledge. This strength is expressed by how easily a consumer identifies the brand whenever he/she is exposed to prompts represented by the brand itself (brand recognition), or by a product category, by the needs met by that category and by brand recall.

Brand image is made up by the bulk of cognitive associations into one's own memory. Such associations have a meaning attributed to the brand by consumers and are a summary of the following:

- product knowledge;
- self-knowledge;
- relation knowledge.

Networks are separate products that offer a unique experience to viewers. Networks differentiate themselves by offering different genres of programming at different times to different demographic groups. In addition to its unique blend of programming, each network also has a distinct logo, jingle, style of promotion, and set of on-air personalities that make up what is known as its interstitial programming.[2] Together, carefully blended programming and interstitials give networks a particular "look and feel." Market researchers often test this concept with viewers, who unfailingly identify networks from images or descriptions.

COMPETITION PRESSURE INCREASE

One of the major impacts from digital technologies is a gradual development of television services in terms of the quality of the broadcast signal as well as in terms of number of television channels available. As pointed out in the previous chapters, through electromagnetic space compression, with the same amount of frequencies needed for an analogic television channel, now one can count upon four to six digital channels, while an even greater number may be available in the future. The new television market based upon the advent of digital technologies entails an increase in the number of available channels and more specifically a wider choice.

First of all, network proliferation generated by new digital technologies allows the single user a greater number of choices. Yet, the average user's viewing time appears to be the same (25.5 hours per week[3]) and channels choice is reduced to a *core group* (seven to nine channels) based upon what users wish to watch, their expectations and interests.

One of the major consequences of this process is that television channels must increasingly have a strong, relevant, and coherent brand to defend their competitive position over time, and be recognized and chosen.

Digital Television Networks Types

Based upon choices made by television networks in terms of differentiation, clusters of viewers reached, and types of programs offered, a number of business areas can be identified in the television industry. A first option is an *undifferentiated strategy* vis-à-vis the audience and types of programs offered. This is the case of the general networks offering general programming made of different types of programs and addressing all categories of audience (for example, BBC and Anglia Television in the UK, RAI and Mediaset in Italy, France 2 and 3 in France).

A second option is that of a *concentrated strategy*: the network focuses on a specific target audience or genre of broadcasts. This means a segmentation according to one or both of these elements as well as a limited choice—*narrow scope*—to a very limited number of segments to serve with ad-hoc programs.

Figure 6.1: Digital Television Channels Breakdown

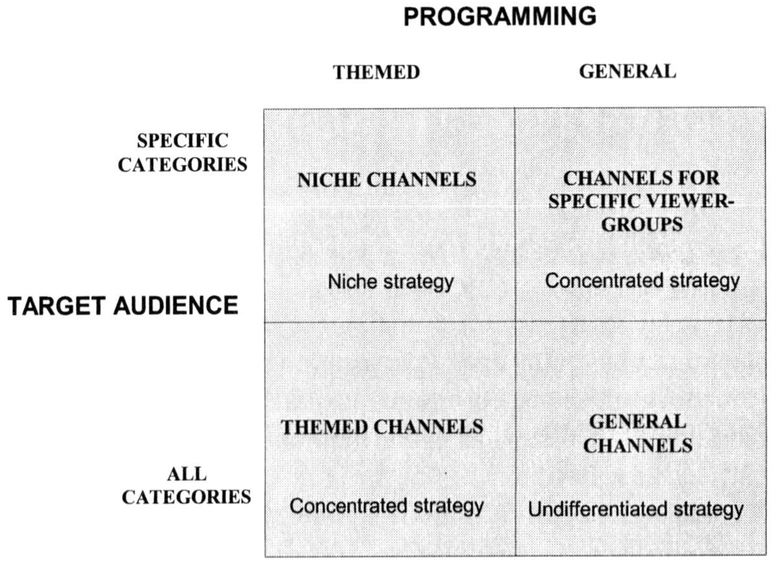

TARGET AUDIENCE	PROGRAMMING	
	THEMED	GENERAL
SPECIFIC CATEGORIES	NICHE CHANNELS — Niche strategy	CHANNELS FOR SPECIFIC VIEWER-GROUPS — Concentrated strategy
ALL CATEGORIES	THEMED CHANNELS — Concentrated strategy	GENERAL CHANNELS — Undifferentiated strategy

If this strategy is pushed forward to the point of identifying a single genre of broadcasts for a single segment, then a *niche strategy* exists. There are several examples of concentrated strategy, such as themed digital channels broadcasting a specific type of program (e.g., National Geographic, BBC News, Disney Channel, MTV, Sky Sport). In the year 2002, in Western Europe 832 digital television channels (66%) out of 1,268 offered themed programming.[4] There are also several examples of channels for specific target viewers offering specifically conceived programs (e.g., Dazed Television and Bravo are targeted to a young audience). A niche strategy is indeed adopted by those channels, having a single programming genre for a specific audience (e.g., Cartoon Network).

Based upon the above, the following types of television channels can be identified:

- *General channels,* both private and state-owned, featuring general programming made by several genres (news, entertainment, films, etc.) and addressing all viewers' categories to satisfy their interests and meet their needs. They target audience maximisation and, for this reason, programming is increasingly homogeneous and convergent, made by programs which can attract the highest number of viewers (a blend of comedy and drama series, movies, sports, news, and special programming).
- *Themed channels* achieve differentiation by filling their entire schedule with programming from a single genre (i.e., Sky News, Sky Sport, RAI News 24, BBC News, MTV, National Geographic). Addressed to all categories of viewers interested in a specific theme, they tend to concentrate on those genres, such as sports, music, news, music, or education, which can horizontally aggregate as many viewers as possible;
- *Channels for specific viewer-groups* which focus their programs on a specific audience segment (Diesel Channel or Dazed Channel in the UK for teen-agers and young viewers, or Nick Jr. and The Disney Channel for children).
- *Niche channels* addressing specific audience segments—they offer a single genre (i.e., Cartoon Network or Nick Jr.).

The television market is characterised by an increasing number of themed channels for specific viewer groups. Many themed channels try to differentiate by adding on to their programming some interactive functionalities (i.e., for music channels, options of music on demand, interactive advertising with purchase of CDs from the channel), or becoming increasingly specialised and focused (i.e., a specific sport, or a single soccer team as in the case of MilanChannel or InterChannel in Italy).

The fragmentation of the audience and the abundance of niche channels cause an increase in the advertising space value of those channels, since the efficacy and quality of the contacts made is enhanced.

A critical success factor for these channels is the ability to pass their values onto their audience to allow for self-identification.

Competitive Mechanisms

The layout adopted to define television businesses can also be used when analysing competition and assessing each single competitor's strategy. Therefore, it is worth analysing each competitor according to the audience groups reached and broadcasts.

A first category of competitors is made of those who define their businesses in practically the same way, reaching the same market segments with the same programming genres. For a general domestic channel, a competitor would be another general domestic channel.

A second category of competitors can also be identified, grouping those channels which define their businesses in a partially different way. Differences may involve segments of viewers reached, or genres of broadcasts.

The third and last category is made up of the new potential *competitors* who, at least in theory, might be granted market access.

Circumstances may of course vary; yet, three fundamental competitive mechanisms can be evidenced.

1. *Crossover Competition*
   Many existing general TV channels enter into new business areas through the launching of new themed television channels (BBC with BBCnews24, RAI with RAInews24, or SKY with Sky News and Sky Sport), or of new channels for niche viewers (in the UK, Nickelodeon UK launched Nick Jr., a channel for children only). They can therefore exploit skills and competitive advantages connected with brand knowledge and loyalty already achieved with the viewers.
2. *Chain Competition*
   Organisations active in other phases of the value chain enter the television market. In particular, the digital convergence process and the develop-

ment of interactive technologies has prompted many organisations specialised in interactivity to enter the television market, such as in the case of Microsoft, Playstation, and Sega. A lot of publishers launch their own channels, such as RCS and Il Sole 24 Ore in Italy.

3. *Intersectorial Competition*

    Organisations belonging to different markets and with no previous television experience set up their television channels which make the most of brand knowledge and image already achieved in other areas (this is the case of Pepsi with Pepsi Music Channel or Sport Crazy Channel, and of Disney with The Disney Channel).

Several reasons exist for the entry of these non-television brands into this business, including:

- low entry barriers in the TV business (especially at the technological level);
- further strengthening of the brand;
- chances to increase advertising money.

Following the present evolution trends, image brand takes on an increasingly central role and must perform a relevant mediation role allowing viewers to understand personality, corporation, and privileged values in doing the business, and to connect both tangible and intangible attributes offered with the socio-psychological benefits they are looking for.

Within this perspective, both TV and non-TV brands existing on the market have a greater advantage during start-up. A strong corporate brand:

- plays a differentiation role;
- represents a strong indicator for basic skills available to an organisation and benefits obtainable by viewers;
- represents a strong credibility and identity element.

Within this new competitive scenario, the following are the most successful brands:

- TV and non-TV brands existing on the market;
- brands having previously accumulated a strong credibility vis-à-vis viewers, also being active in different businesses (i.e., Nike and Virgin), or

which have distinct programming features (BBC and SKY) or an interactive level offered (Sega or Playstation);
- the first entries.

LAUNCHING OF A THEMED CHANNEL ON THE ITALIAN TELEVISION MARKET: THE DISNEY CHANNEL CASE

A significant example of the entry of a non-television brand into the market is the themed Disney Channel case, the first example of Italian "*brand television.*"

Disney, upon entering this new market, had a strong brand known and loved by Italians for more than 70 years. Disney's target was to have a channel which could strengthen and be coherent with its already established brand quality.

The decision to start with a themed channel required a careful entry strategy. First, an accurate business plan was developed to study the market, competitors, advertising strategy, distribution issues, and growth prospects. Creativity and product quality were the staple concepts to be immediately successful and achieve viewers' satisfaction from scratch.

The Disney Channel operation started with a number of specific research efforts on the television market in order to understand more of the relationship that both children and their parents bear with television, family habits, and product expectations. Behavioural changes—as well as changes in habits—in the course of the day were also investigated. Viewers' psychology was carefully investigated as a key factor to define a correct positioning vis-à-vis a niche audience.

Indeed, television for an audience niche is based upon understanding the viewers who are the recipients of what is being offered.

The following step was useful to know the competitive scenario, while the third phase of the strategy was devoted to channel positioning.

Most of the television brands may be broken down into the following three categories:

- those based upon a genre; i.e., single-themed channels such as CNN or Cartoon Network;

- those addressing a target, such as Lifetime Television (channel devoted to soap operas for women) or Women Television;
- those developed around a developer, such as in the case of Disney's brand defined by its inventor Mr. Walt Disney.

Positioning is one of the strong elements of the *brand television* since it guides the whole business. In the case of The Disney Channel, positioning definition was guided by the brand. The target audience of choice was the family with programs especially conceived for it. Identity and image of quality were also the reference point to select human resources for The Disney Channel.

Following positioning definition, target audience, and having set up a network of human resources as well as the corporate organisation, the time came to approach the issue of management. Disney's programs are not interrupted by commercials and are focused on children and families.

The success factor for The Disney Channel is the *promo,* whose target is not that of raising the interest for a given broadcast, but rather telling a story contributing to increase brand loyalty. A relevant factor is also a promo jingle, where focus is placed on music, synchrony with the lyrics, and the content of the story. A decisive contribution also comes from the "*idents*" which are made of different shapes visually recalling The Disney Channel logo.

Therefore, the marketing function plays a central role for a channel like this one and must promote brand values through "on air" and "off air" elements to increase viewers' loyalty over time.

In the light of the above example, a successful brand extension strategy can be turned into an improved brand image, thus increasing loyalty and consolidating relationships with customers.

Figure 6.2: Launching of a Themed Channel: The Disney Channel

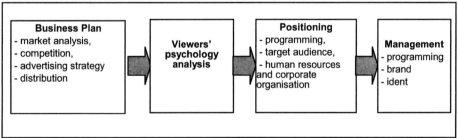

It is worth pointing out that the concept of brand image can be broken down into the following fundamental components[5] (Howard, 1977: 45-46; Busacca & Troilo, 1992: 75-76; Busacca, 1995):

- *identification component:* related to distinctive features and characterising items (name, logo, etc.);
- *evaluation component:* specific values and connotations (emotional and psychological factors);
- loyalty component.

VIEWERS' COGNITIVE SYSTEM ANALYSIS PARAMETERS

In order to understand the way processes triggering loyalty resources by the TV broadcaster work, one must analyse in detail the rationale connected with viewers' behaviours in making use of a television program.

The fruition process and viewers' analysis become central in the process of achieving an understanding about a suitable management of the relationships with the market. More specifically, focus must first of all be placed on the development of this process by highlighting the elements influencing its complexity and then focus on the following:

- motivational system, given that the nature of the benefits achievable within the programs depends upon it;
- perceptive system which, through categorising data of different nature from different information sources, provides an orientation to comparison processes as well as on suitability of choices made;
- sequences and evaluation procedures essential to create the above approaches both before and after viewing experiences.

A user's cognitive process can be broken down into a series of logically sequential phases[6] defining its layout as well as its morphologic and dynamic complexity (Figure 6.3):

- perception of a general need (the need to watch TV), or a specific need (a specific genre or program);

Figure 6.3: User's Cognitive Process Phases

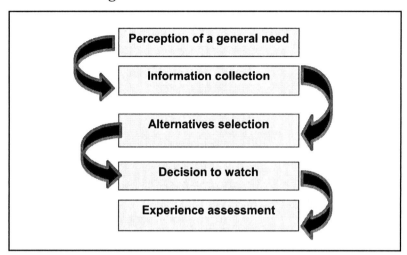

- information production suitable to give an orientation to comparisons and choices;
- assessment on optional offerings on the market;
- decision to watch;
- assessment on the basis of one's own watching experience of the matching between expectations developed during the previous phases and the actual benefits offered by the program of choice.

Before analysing each single phase, one needs to know that the program and the channel play different roles within the different phases of the user's cognitive process, and one needs to start from the television viewing *basics* in order to fully understand this role.

An individual stays tuned and enjoys a specific program in order to satisfy a generic need for television watching or a specific need to watch a specific program ("watching date").

During this first phase, the user's goal is his/her research for a program, or a much looked forward time to deciding how to spend his/her time. During this first phase, the program is the final element of choice. The relationship with the channel, therefore, is limited to simply *browsing through,* and no specific attention is being paid to it.

From a study carried out on the British market,[7] viewers watch an average of 30 to 40 programs a week; yet, only six or seven of them are truly "watching dates."[8] Furthermore, viewers can express an opinion on the program watched; yet, they have trouble remembering the channel where the program of choice was broadcast, except for a few specific times (between 6 p.m. and 11 p.m.), and some specific genres (news) where they can accurately tell the channel and broadcasting times.

The assessment of optional choices (third phase of the cognitive process) is instead made based upon the channel.

In a television environment characterised by hyper-competition, viewers have problems checking out all the channels which indeed are offering options (**awareness aggregate**), thus driving the choice process. Therefore, a buffering process is activated only among a given number of channels visited regularly (**evoked aggregate**).

On average, each viewer makes a choice between seven to nine channels which are being watched on a regular basis. This set of channels is made up of those meeting the viewer's expectations in terms of genre type offered, type of programming, and ability to satisfy a number of perceived needs by viewers in the different time brackets. The channel goal is, therefore, viewers' loyalty development, as well as strong reputation achievement in terms of being a competitive source.

The evoked aggregate by channel brands has a steady trend, and previous learning is the most meaningful influencing factor in determining behavioural orientations.[9] The danger for those channels which are not part of the set of favourites (**negative group**) is to be connected in the viewers' minds with past prejudices developed by them.

Figure 6.4: Options Selection

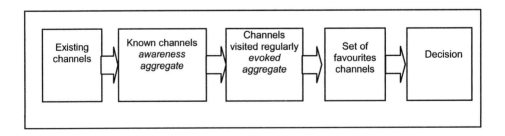

CHANNEL-WATCHING MOTIVATION TRIGGER

Many factors are responsible for the triggering of cognitive processes at the basis for the use of a specific channel or program, and assessment of the satisfaction level achieved[10]:

- Exposure Variables:
  - **titles** broadcast following advertising breaks to convey useful information in a non-intrusive fashion for the purpose of setting up a long-lasting relationship with users and leading them to place the channel within their choices;
  - **EPG Electronic Program Guide** or Electronic Guide allows users to achieve the following:
    - find titles and timetables of the events in compliance with specific criteria (selectable), such as time, program type, subjects, promotional adverts;
    - automatic access to the selected program;
    - set terminal for recording purposes;
    - send Pay-Per-View request;
    - access further information.
- Influencing Variables:
  - **program title** as a means to attract audience—a good title raises feelings (suspense, tension, interest, curiosity) and gives an idea to viewers about what they can get in exchange for their attention;
  - **channel logo** is as relevant as the program since it is a clear-cut invitation and conveys the key message to the market segment addressed.[11]
- *Single User's Inclinations:* a new channel is often found through **zapping,** which has no basic rationale but rather is a "casual" search for a program, with no specific attention to the channel broadcasting the program. Zapping is a totally *random* exercise and is not guided by the channel but by the viewer. In a multi-channel environment, a great number of channels and similar options are available.

Audience loyalty to the channel depends upon the chance of finding one's own favourite programs. A positive assessment based upon a watching experience on the matching between expectations from previous stages and benefits actually found in the program of choice allows the viewer to have a positive channel experience which develops and gets consolidated through repeated contacts over time. Therefore, it is important that values to be communicated by the brand and channel identity be daily reflected by the program schedule. The decision of tuning into a channel is only being made in the event the viewer had a positive past channel experience and is aware that his/her expectations can be met by a specific brand.

Meeting audience expectations means developing the viewers' confidence vis-à-vis a specific brand (channel); all this allows for channel loyalty growth. Development of channel loyalty is based upon the following four criteria:

1. *Knowledge:* The channel brand must be known to viewers who must be able to understand its key values and differences from competitors in order to be able to select it against competing channels.
2. *Experience:* Each TV program selected by a viewer on a given channel must meet, within a competitive environment, the user's expectations so that through a positive experience he/she can once again select that specific channel.
3. *Relationship:* Regular and frequent contacts between the consumer and the brand help build up a relationship so that, also in a competitive environment, the viewer is able to recognize the channel brand and can put it in the core group of his/her selected channels.
4. *Trust:* If the relationship between channel and viewer is strong and frequent, and the user has made a positive channel experience, trust is built originating in the viewer an expectation whereby that specific channel brand will keep a promise and confirm his/her positive past experience.

When a viewer has a positive experience in all four phases, channel brand loyalty is built and made stronger.

TELEVISION BRAND

After having outlined market competitive mechanisms as well as the cognitive process guiding users in making their watching choices, a better understanding of the meaning of a channel brand must be achieved. Not only does the brand possess an intrinsic relational character, but it also represents one of the key loyalty resources the broadcaster can count on in the pursuit of long-term growth within a situation of growing competition.

A brand is a product that is differentiated from its competition by means of design, name, mark, imagery, or a combination of any or all these.[12] Firms trying to sell their products and build loyalty with their customers in a crowded field of competitors use branding to distinguish their products.

At a minimum, brands identify products as different. But, brands can go further, to identify products with positive attributes. Well-liked brands can give consumers trust and confidence in products. In this way, consumers pressed for time can quickly make difficult choices by selecting a known brand with an established reputation.

Brand equity is that something extra, beyond the value of the physical asset of the product, that gives the product its commercial appeal. Creating a successful brand is quite difficult. Branding requires significant and repeated investment.

Brand is no longer a distinctive element of large industrial corporations; it represents a fundamental element within marketing policies of every type of enterprise,[13] inclusive of TV broadcasters. More specifically, in the television business, branding is more common in packaging than in production or distribution. In fact, in 1998, two of the first 50 global brands turned out to be TV channels: CNN (33rd) and the BBC (50th).

As it happens for services and consumption products, branding is becoming an increasingly interesting and wider element, also with reference to the media industry (TV, press, radio, etc.). The broadcast networks are now trying to build on their brands to retain and increase viewership. Instead of using their valuable time between programming to sell advertisers' products, they are using it to promote themselves.[14] The time is used to promote both individual shows and entire evening schedules.

With reference to television, the following three brand levels can be devised:

- *corporate* with reference to the broadcaster (i.e., ABC, CBS, BBC);
- *channel* (i.e., The Disney Channel);
- *program* (i.e., Beverly Hills 90210).

Each of these brands has its specific lifecycle (Figure 6.5).

While "corporate" brand gets consolidated over time, at the channel level a brand development stage can be broken down into the four following different phases: launch, growth, maturity, and decline. Single programs definitely have shorter lifecycles.

The countless number of channels and programs offered forces television networks to consider branding as one of the relevant factors within a new competitive concept. Brand indeed allows for the following:

Figure 6.5: Brand Lifecycle

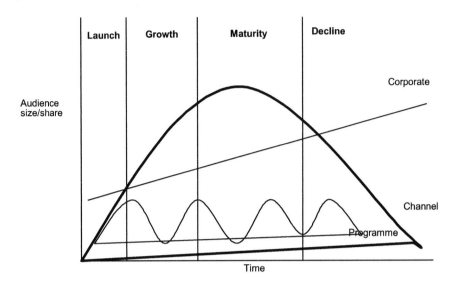

- differentiation vis-à-vis competitors as a primary element within corporate market policies;
- immediate recognition of the channel and program by the viewer;
- an important reference point, a form of help, an active guarantee for the user (loyalty);
- an immediate recognition of the quality level of the offer.

Last but not least, it may be useful to bear in mind that correct management and ongoing development of a brand policy allows the broadcaster to achieve and maintain audience loyalty and, as a consequence, make viewers an "asset" through repetitive and loyalty-oriented behaviours (Busacca, 1994). The four factors mentioned above (differentiation, recognition, quality, and loyalty) seem to be central in order to obtain a competitive advantage.

BRAND COMMUNICATION TOOLS AVAILABLE TO DIGITAL TELEVISION CHANNELS

Tools available to digital television channels for brand communication purposes may either be *on-air* or *off air*.

On-air tools are the programming schedule offered, that is the bulk of programs, style, atmosphere, and the energy reflecting what the brand needs to communicate. Such tools, day in day out, contribute to make the relationship between a channel and its viewers stronger.

At a time when the branding experience is broadcast *"on air,"* it can be switched into an outer *"off-air"* environment, making sure that key values do not get lost or altered, but are rather made stronger. For the purpose of developing an *"off-air"* awareness, it is important that the following be defined by the television channel:

- Who is to be addressed (target audience)?
- Why is that specific group of viewers being targeted?
- What are the previous expectations of the target audience?
- Where can the target audience be reached?
- What is the best possible way to convey an actual "branding experience" to a specific market segment?

The choice of the available *off-air* tools depends upon the type of brand and its very nature. The following are the tools to make a stronger *"off-air"* branding awareness:

- *Live Interactive Road Shows* allow for brand accessibility and "interaction" with viewers. This is a tool which may not be relevant and appropriate for all types of brands, but it works very well with television brands (this communication tool is used by Tele+ in Italy and SHS Multimedia in Europe).
- *Press* (magazines/press) may strictly be oriented to the target audience. Very good from an economic standpoint.
- *Outdoor* has a strong impact according to sizes and positioning of the communication tool. As it happens with the press, it is a static medium and has greater difficulty conveying an actual branding experience.
- *Direct Mail* can strictly be oriented to the target audience with minimum losses and great economic impact; yet, it runs the risk of being considered as one of the many "advertising brochures" and not be read.
- *Movies* give a strong branding experience; yet, one needs to ensure that key values connected to the channel by viewers may be present within the creative performance.
- *Promotional Business* provides access to a wide audience.

- *Other Tools* include merchandise, consumption products, entertainment, websites and sponsorships.

In communicating a brand, both *on air* as well as *off air,* the channel must greatly focus on what must be communicated and how to do it in order to be coherent with promises made and expectations raised in the viewers' minds.

INTERACTIVE PORTAL TOOLS AVAILABLE TO INCREASE VIEWER'S LOYALTY

The development of interactive television entails a central role of the portal whose target is to supply contents and interactive applications, as well as to develop a long-lasting relationship with viewers.

A portal is an entryway, a navigation aid, a communication device, and a revenue opportunity. As a navigator, it must have easy and powerful search capabilities, and a visually appealing interface.

In order to exist in a meaningful way, the TV portal must provide value, either entertainment or informational value. To understand how interactive television is going to be accepted by the consumer, you should start by looking at the historical behavior of the television experience.

Television is a group and passive experience. People turn to their TV to be entertained, and programmers are not looking for more competition. For a long time to come, people are going to turn on their television sets to "watch TV." In this perspective the cable or satellite and terrestrial operator, who wants to get additional revenues or t-commerce from interactive television, has to consider that interactive television must be relevant to the viewing experience. Control of the TV portal represents a critical aspect to be considered by the MSO in order to take the opportunity to share the control of a home page.

The second important aspect is represented by the shape of the portal. In the most important TV portals in the U.S., full screen overlays with a video window or partially overlays on top of the video. It's a fundamental belief that the primary access to the homepage occurs during breaks in programming and that the first screen is the most important tool for introducing and promoting (marketing) new interactive television offerings.

What does this mean to the cable, satellite, or terrestrial operator who wants to offer new opportunities for content and interactive television within the portal? The portal must generate traffic and must have the ability to promote.

In interactive television, if it is not a significant subscription service, it contains advertising. In this case it is important to define the correct proportion between content and advertising: up to 30% of the screen can be advertising or promotion without being offensive, provided that the rest of the screen brings value.

The following are services offered within a homepage:

- IPG (Interactive Program Guide);
- entertainment-related sites;
- shopping;
- non-traditional Video-On-Demand (VOD) content;
- email and/or chat;
- local content.

Offered beyond the homepage are:

- information or utility services, such as news, weather forecasts, sports, stocks;
- applications that overlay programming with relevant information or services.

The TV portals are very dependent on technology and intellectual property, including:

- set top boxes that have processing power and memory;
- server scalability and connectivity;
- bandwidth.

The MSO is the only one guaranteed to survive. As application providers, the homepage, the IPG (Interactive Program Guide), and others must all provide value to survive.

CONCLUSIONS

How can one create a channel experience by getting hold of viewers to the detriment of one's own competitors? How can one enhance the value of access to an interactive television portal by viewers? The results achieved—those being the outcome of interviews of some digital television channels and interactive portals people—highlight the central role of marketing policies adopted.

The marketing function must control marketing services offered, the communication policy on the air, and advertising, as well as research and planning. For the purpose of making the marketing strategy stronger, new off-air communication tools must be introduced, a steady *breakthrough* level must be achieved, and a focus on other media, such as the press and radio, must be accomplished.

Independent of the vehicle of choice, brand personality, values, and philosophy must be reflected in the contents offered by trying to meet the expectations raised in the viewers' minds, based upon promises made and previous experiences. A positive channel awareness and experience may allow the viewer to increase his/her confidence in the brand, select a channel again, or have access again to the interactive portal.

ENDNOTES

[1] Valdani, E. & Busacca, B. (1999). *Customer Base View*. Finanza Marketing e Produzione, Vol. 2, pp. 95-131.

[2] On this topic see also Todreas, T.M. (1999). *Value Creation and Branding in Television's Digital Age*. Westport, CN: Quorum Books, p. 175.

[3] Source: BBC, 2000.

[4] New Media&TV-Lab. (2002). *Rapporto Annuale: La Televisione Digitale in Europa*. Research Report, New Media&TV-Lab, I-LAB Research Center on Digital Economy, Bocconi University, Milan, Italy.

[5] Busacca, B. (1995). Le strategie di brand extension: l'attivazione del valore-potenzialità della marca. In Adams, P., Bertoli, G., Busacca, B., Gnecchi, M., Mazzei, R., Verona, G. & Vicari, S. (Eds.), *Brand Equity: Il Potenziale Generativo Della Fiducia*. Milan, Italy: EGEA, pp. 157-198.

[6] Busacca, B. (1994). *Le Risorse di Fiducia dell'Impresa*. Turin, Italy: UTET, pp. 73-74.

[7] Source: Broadcast Research Ltd., 1998.

[8] Cirone, N. (1999). Managing director broadcast strategy. Speech presented at the conference, *Brand Strategies for TV Channels*, in London, April 14.

9   Aaker (1991: 61-62) identifies four levels of brand renown: 1) unaware of brand; 2) brand recognition; 3) brand recall; 4) top of mind.

10 Lazarsfeld, 1934, 1935; Lazarsfeld & Rosemberg, 1955.

11 In 1996 Tele+, the Italian satellite platform, carried out a careful analysis of its channel logo considering it as one of the main brand communication tools.

12 Todreas, T. M. (1999). *Value Creation and Branding in Television's Digital Age.* Westport, CN: Quorum Books, pp.172-173.

13 Fiocca, R. & Corvi, E. (1996). *Comunicazione e Valore Nelle Relazioni d'Impresa.* Milan, Italy: EGEA.

14 Todreas, T. M. (1999). *Value Creation and Branding in Television's Digital Age.* Westport, CN: Quorum Books, p. 177.

Chapter VII

The Critical Role of Content Media Management*

INTRODUCTION

The advent of digitalisation is providing big opportunities which are changing the shape of the broadcasting industry. New business models and revenue opportunities based on digital capabilities are emerging. Digital technology provides digital media companies with the opportunity to:

- provide new products and cross-media experiences aimed at individual consumers and like-minded groups;
- provide premium services that give viewers access to greater and more personalized content;
- change the television advertising model from mass media to targeted advertising (market and advertise on one-to-one basis direct to consumer);
- expand from a traditional push model to a pull-model for product distribution.

---

* *Earlier version of this paper was presented at IRMA 2001 Conference (International Resource Management Association) in Toronto (Canada), May 2001, and at several conferences sponsored by SHS Multimedia in London, Paris, and Madrid, November 2001. The author wishes to acknowledge the input of participants at these conferences and the company Mediaset (in Italy) for the important experienced suggestions.*

It has often been cited that one of the biggest problems with the convergence of diverse cultures is the ability to speak a common language. As we enter the 21st Century, we are learning that the convergence of the formerly isolated industries that deliver entertainment and information to the masses requires the development of a common language for digital communications.

The development of a universal language of ones and zeros is a critical enabling factor for the convergence of formerly diverse media. Likewise the inter-working of all digital communications media enables a wide range of new applications that are being lumped under the broad umbrella of "digital." However, working with digital data does not guarantee that one device can talk to another, any more than the use of a common 26-letter alphabet guarantees that a Frenchman can communicate with an Englishman.

The new society of information is thus facing a major issue, due to the growing mass of multimedia data which every day are created, managed, and distributed, by crossing over any borders and barriers.

But convergence is not solely a technological matter: it's a brand new life and working style, as it foresees new services and opportunities to implement industry productivity and competitiveness on the market: media, information technology, and telecommunications are then to take advantage of new products and platforms to become part of a unique global network.

In fact, the most impressive revolution in television over the past few years has been digitalization. Digital communications are the first and essential step to the convergence of information and telecommunications technology where traditional media, once clearly distinct and independent from each other, meet in the new land of interactivity and multimedia.

Archiving, accessing, managing, and security of digital content assets become basic requirements in the everyday life of multimedia creators and providers. This ability can effectively be defined as a company's requirements for content management. Content management is becoming a core capability at the heart of all media and entertainment enterprises, supporting everything from content creation and acquisition through to personalisation and customer relationship management.

The purpose of this chapter is to understand the advantages offered by the implementation of content management for a digital media provider and how it integrates the business process value chain changing the nature of a company's assets and products.

The chapter, after giving a definition of content management, focuses on how digitally enabled processes, fed by content management, integrate the business and allow it to evolve, transform, and realise new strategies and opportunities based on enabled capabilities. The chapter then considers the content management strategy adopted by digital television companies and changes required to business process, organization, system, and culture, across the enterprise. The practical implementation of the concept will be illustrated on the example of a case study.

DEFINITION OF CONTENT MANAGEMENT

Content management can be defined according to three views[1] (see Figure 7.1):

- **Enterprise-wide** content management means capturing, storing, managing, exploiting, and protecting a company's digital assets throughout their lifecycle within the enterprise.
- **Digital asset management** is the combination of business processes and systems that allow the business to capture, store, and manage commercially exploitable digital assets and associated data in common formats for use across the organisation.
- **Channel content management** is the combination of systems and processes that enable the creation/capture, formatting, and storage of trusted channel content, control of when and how content is released,

Figure 7.1: Content Management

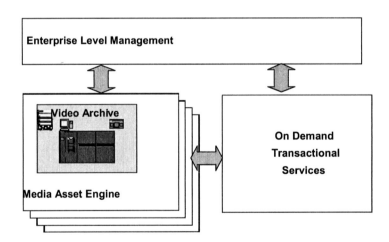

Source: Adapted from Sony Broadcast and Professional Europe, 2000

where and when it gets used, and how long it remains there. This covers text, graphics, multimedia, and applets across the Web, 3G mobile, digital television, etc.

Content management software and technical standards are evolving. Most current content management products and services are focused on digital asset management or channel content management, and the market is constantly changing as new companies appear and existing companies merge.

The requirements for digital asset management and channel content management products are converging[2] as:

- digital asset management vendors and content creators realise that valuable marketing materials (sound bites, text, clips, logos, photos, ads) and product specifications can and should be captured during the creation process and used as channel content;
- channel content vendors and developers realise that channel content has to be driven from and integrated with the business value chain, to ensure that channel content is accurate, consistent, and up to date.

Implementing content management is a system integration activity.

The following analysis focuses on three critical areas of content management (Figure 7.2):

- *Media Asset Management (MAM)* means the management of all tangible assets (audio, video, etc.), from conception to completion throughout the production chain.
- *Archive Management* is concerned with the management of materials that have completed the production process.
- *Delivery Management*.

Media asset management is the central function when the finished product is bought by broadcaster. The digitalisation of the television signal and the storage of the materials that have been purchased allow broadcasters to build metadata and facilitate research of the material.

Broadcasters can, in fact, make a selection of audio and visual contents after they have purchased them, recovering, reusing, and offering them again on different platforms. This process implies a cost reduction and a reduction of the risk of loss (the same contents can be used more than once in different ways), as well as the generation of greater profits.

160 Pagani

The storage of the digital signal and the ability to recover it allow the sales force to develop new services and to sell these in new markets or platforms. Distribution plays a critical role in supplying digitalised contents through

Figure 7.2: The Critical Areas of Content Management

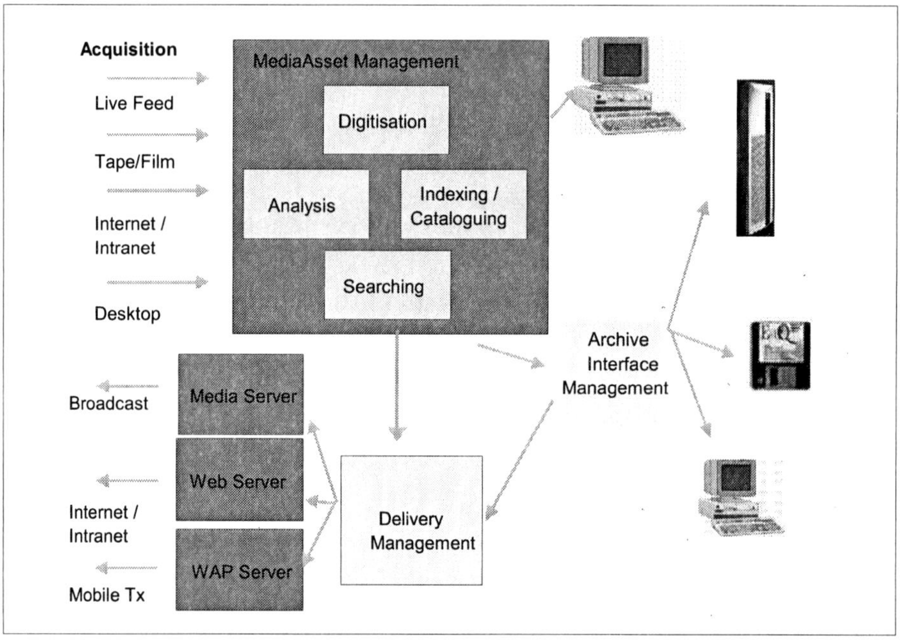

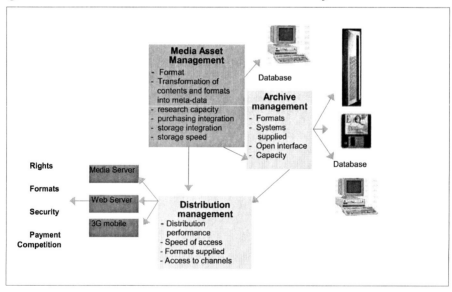

Figure 7.3: Success Factors in the Critical Areas of Business

different platforms and formats (TV, radio, Web, 3G mobile). The key factors are:

- distribution performance;
- fast access to the stored contents;
- the possibility to create different formats;
- access to channels.

MANAGERIAL IMPLICATION FOR THE DIGITAL BROADCASTER

Media asset management and the construction of a digital archive naturally involve the whole broadcaster in the redefinition of strategies, changes to the business model, and organisational changes.

What are the managerial implications of the application of content management for digital broadcasters? In the first place it seems worthwhile showing how the value chain[3] for broadcasters changes.

The traditional value chain in the entertainment and media industry is focused on the release of physical products to a pre-planned schedule. It's often a disjointed, manually intensive process involving the physical movement of many masters and production of a large amount of support material, which is often lost once the product has been released.

Figure 7.4: The New Value Chain for Digital Broadcaster

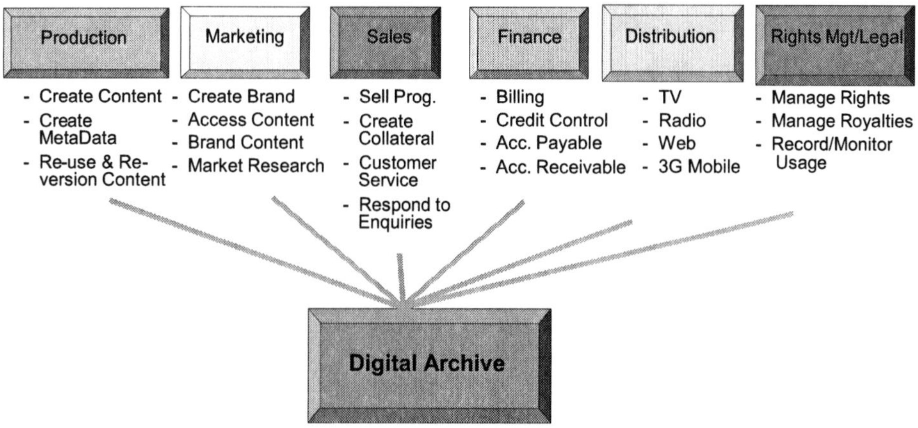

Production	Marketing	Sales	Finance	Distribution	Rights Mgt/Legal
- Create Content	- Create Brand	- Sell Prog.	- Billing	- TV	- Manage Rights
- Create MetaData	- Access Content	- Create Collateral	- Credit Control	- Radio	- Manage Royalties
- Re-use & Re-version Content	- Brand Content	- Customer Service	- Acc. Payable	- Web	- Record/Monitor Usage
	- Market Research	- Respond to Enquiries	- Acc. Receivable	- 3G Mobile	

All feeding into: **Digital Archive**

Figure 7.5: Processes More Impacted by Content Management

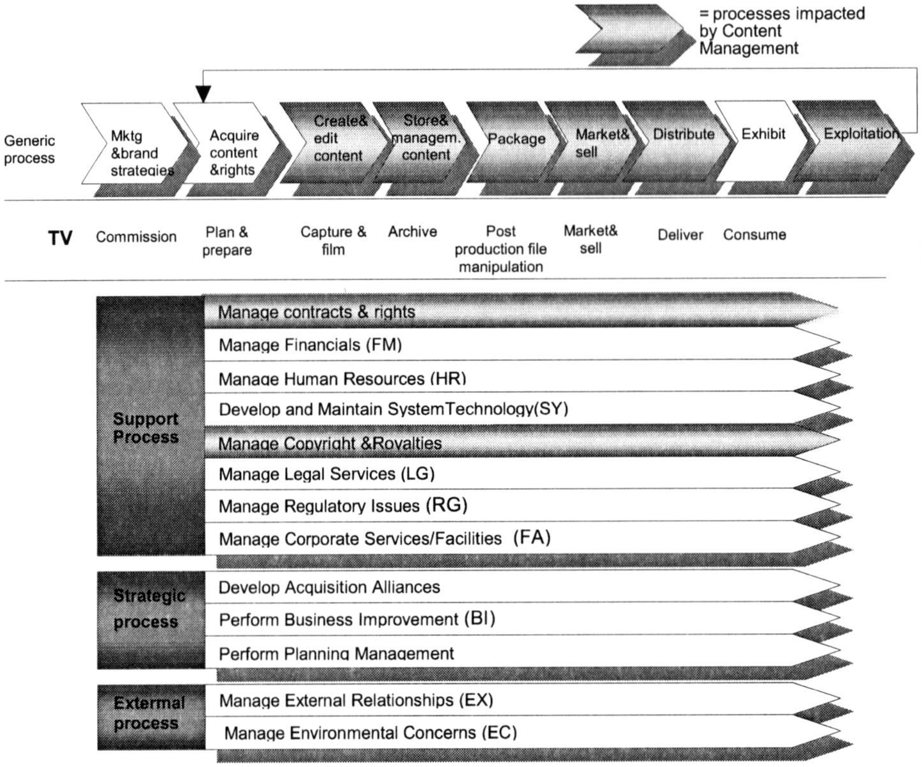

Source: Elaboration from Pricewaterhouse Coopers, 2000

The implementation of content management integrates the business process value chain and fundamentally changes the nature of a company's assets and products (Figure 7.5).

Assets captured digitally can be used in the assembly, marketing, sales, or promotion of a product or service (i.e., video, text, audio, software, graphics, logos, etc.)

A combination of digital assets can be packaged together as digital or physical products to address a market opportunity regardless of the media type (audio, text, and video compilations).

Integration provides a pool of re-usable digital assets, products, and association data, in common and easily transmitted formats. These not only enhance existing channels but also open, previously unavailable opportunities for new products and channels.

Aside from the creation of new opportunities, media asset management makes it possible to offer services that provide new and additional revenue (e-commerce, advertising, Pay-Per-View, and premium services) to portal operators and content creators. Under this concept, video and audio can be pulled over telecommunications lines and accessed using Internet-based applications. Digital portals provide the best means for searching, selecting, purchasing, and using content by organising access to the available materials in a commercially viable manner.

The construction of an archive provides the broadcaster with many benefits:

- check the material that is owned;
- facilitate research and access to selected material (metadata extraction);
- re-purpose contents and information to create new services and generate new profit sources (distribution, sales, contents licenses in new markets);
- protection of contents from deterioration;
- standardisation.

Media asset management provides a dramatic improvement of asset supply chain (Figure 7.6) reduction of production and distribution costs and opportunity loss, and it makes it possible to generate new profit.

Figure 7.6: The Asset Supply Chain

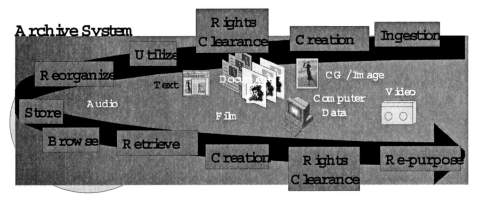

Figure 7.7: Primary Barriers to Implementation

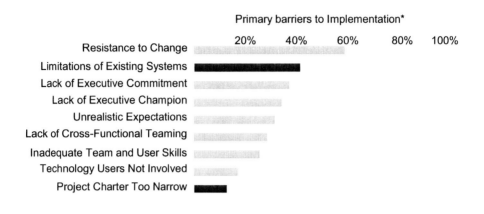

Source: Information Week: Survey of 400 CIOs

There are some pitfalls that broadcaster management has to consider, and these can be summarised as follows:

1. selection of technology;
2. analysis of process changes;
3. definition of the budget;
4. disaster recovery;
5. migration to new technology;
6. staff willingness to embrace new technology and practices.

The cost of technology selection includes business case preparation, bid costs, asset management selection (one or more pilot system installs), and solution selection criteria (existing infrastructure).

Business process implies documenting existing process, defining the goals of the system, gap-analysis followed by re-engineering of new business process. Once a new process is in place, it will need adjustments and fine tuning.

The budget has to consider:

- *human resources expenses* such as IT training, system training, running cost of new infrastructure, resistance to change, re-deployment;

Figure 7.8: The Impact of Media Asset Management on the Other Business Functions

[Figure 7.8 content:

Organisation: New processes, Flow of competencies, Resistance to change, Organisational change

Production: News, Post production, Purchasing, Content creation

Infrastructure: Networks, OS, CPU, Database, ISV, Legal Systems, Business, Availability, Flexibility

Media Integration: Media Quality, Criteria Selection, Commercial Development

Metadata: Which tools, What costs, What level of depth, What standards

Services: B2B, DVB, Portal on line, IP, Internet, DVD CD, Intranet

Strategic Planning: Competitive Advantage, Merge & Acquisition, New business models, New value chain, Technological impact, Standardisation impact, What threats

Legal Systems: Access, Geographical area, Single use not negotiated]

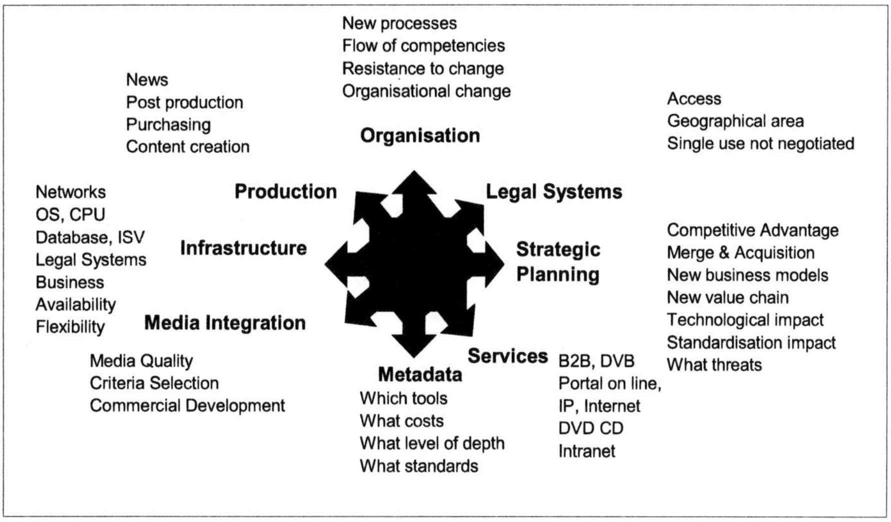

- *system costs* such as hardware and software, customisation and consultancy services, maintenance and support services, staff training, and provision of first-line support.

Media asset management generates an impact on all other business functions, such as organisation, legal systems, strategic planning, services, metadata, media integration, infrastructure, and production as shown in Figure 7.8. All the company is impacted by the digitalisation process and by the adoption of media asset management.

BUSINESS EVOLUTION THROUGH DIGITAL ENABLEMENT

Despite that many well-known discussions of strategy invoke the image that strategy is a course of action consciously deliberated by top management (e.g., Chandler, 1962; Andrews, 1971) or an analytical exercise undertaken by staff strategists (e.g., Ansoff, 1965; Porter 1980), descriptive analysis of the complexity of real organizational phenomena challenges such simplified conceptualization (e.g., Allison, 1971). An explicit recognition of inherent

Figure 7.9: The Four Stages of the Analysis

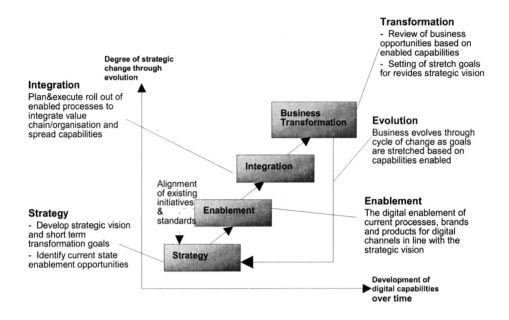

interorganizational complexities, often described as "possible goal incongruence," "information asymmetry," and "organizational politics" (e.g., Barnard, 1938; Simon, 1945; Cyert & March, 1963; Croizier, 1964), as well as "unpredictable" and "uncontrollable" environments (e.g., Schumpeter, 1934; Nelson & Winter, 1982; Thompson, 1967; Pfeffer & Salancik, 1978; Miles, 1982), has led some strategic management scholars to describe how strategy is actually formed instead of prescribing what it should be.

Findings from the empirical studies suggest that rather than seeing the change required as a wholesale revolution, broadcasters should consider taking an evolutionary approach and develop their businesses and capabilities incrementally over time.

To do this they can start by identifying content that is of value in the current process and capturing it at the point of creation, then digitally enabling current business processes over time to use it. This approach means that rather than looking to implement enterprise-wide content management from day one, companies can develop their digital capabilities at their own pace and in tune with changes in the marketplace and advances in technology.

Four key stages describing business evolution through digital enablement are considered:

1. strategy;
2. enablement;
3. integration;
4. transformation.

Strategy

The strategy defines a vision of the space the business wants to occupy in the digital environment, and it identifies a set of short-term goals (over a period of weeks rather than months) and the capabilities required to reach them.

The questions that need to be answered include[4]:

- Who are the target customers? (e.g., traditional customers, B2B, B2C, geographies, demographics);
- What are the target products and channels? (e.g. electronic mail order, digital download, digital streaming);
- What capabilities will you need? (e.g., online payment processing, digital fulfillment).

Answering these and other relevant questions will provide a vision for the business which defines the goals, the capabilities the business already has, and those it must develop in order to reach those goals.

Enablement

This stage involves reviewing the current business model to identify processes that are amenable to digitisation. A review of the present processes will allow the company to work out what it needs in order to transfer and transform current products, brands, and processes into a digital context within its overall vision.

One or more processes in a discrete area of the business can then be selected, so that the changes and standards required can be safely defined and refined before roll out across the organisation. For example, a TV broadcaster may digitally enable a discrete production process (say news and current affairs) before rolling out to all other program departments.

Integration

The aim of this stage is to roll out the digitally enabled processes and then to bind them all together to ensure consistent access to digital content, data, and systems. This is a two-pronged process:

1) integrating the processes in the current value chain as they are digitally enabled;
2) integrating areas of the organization as the enabled processes are rolled out.

The integration process should be planned in line with the overall vision and could be tackled in order of process, organization, brand, or product.

Transformation

The proceeding three stages will create new capabilities that must themselves be reintegrated into the business. Understanding the newly acquired capabilities will create opportunities for new enhanced products, channels, brand strategies, markets and communities, partnerships, and alliances.

The aim of the transformation stage is the identification and classification of these opportunities so that they can be integrated into the strategic vision and goals of the business. Therefore, after each transformation stage, the vision and goals set out in the strategy stage are revised, prior to the next cycle of enablement and integration. This process takes place on a continual basis but at a rate dictated by the company, thus avoiding the hefty impact of an enterprise-wide change program.

THE MIGRATION TOWARD DIGITAL TV

Evolution or revolution? How do broadcasters perceive migration toward digital TV?

Gradual shifting to digital television may be considered by a broadcaster as a successive and incremental step featuring a standard evolution process (as pointed out in the paragraph above); yet, one should consider that digitalisation in content production, media asset management, and distribution of the television signal represent a real revolution process inside the broadcaster. This

Table 7.1: How Broadcasters Perceive Migration Toward Digital TV

EVOLUTION	vs.	REVOLUTION
smooth and natural process		new cultural approach
is part of technological progress		process re-engineering
is a convergence consequence	Implies	claims for new competencies
		change of user perpectives
		different knowledge base
		leads to the convergence

Source: Adapted from Mediaset, 2001

migration indeed implies a new cultural approach, new process re-engineering, claims for new competencies, and change of user perspectives (Table 7.1).

Migration toward digital TV can be perceived by broadcasters as an evolution process (Table 7.2) which implies:

- smooth and natural process;
- technological progress (from analog to digital);
- convergence: a market trend.

Table 7.2: The Evolution Perspective

EVOLUTION
smooth and natural process
• is in the human being, everything evolves smoothly and naturally
• there is no apparent discontinuity with the past
technological progress *(from analog to digital)*
• *it is expected.., no one can stop it.*
• *helps make things better, apparently improves operations but not necessarily processes*
convergence: a market trend
. *means common understanding, reduce costs, increase competition*
• *make easier interoperability and distribution*
. *represent new opportunities*

Source: Adapted from Mediaset, 2001

Table 7.3: The Revolution Perspective

REVOLUTION
new cultural approach • ..to be real effective in digital domain, change your mind • ..do not only apply technology, re-invent the production process itself
claims for new competencies • ..old fashion competencies don't really apply to digital TV domain • ..the gap to be filled out is quite evident
change of user perspectives • ..the contest moves from hyph driven manual operations toward fully automated software processes chains
different knowledge base • ..as digital TV comes, TV will not be TV anymore • ..in the digital TV era, TV business will transform

Source: Adapted from Mediaset, 2001

It can also be perceived as a revolution process which implies deep changes (Table 7.3):

- new cultural approach;
- claims for new competencies;
- change of user perspectives;
- different knowledge base.

THE MEDIASET EXPERIENCE IN ITALY

The analysis of Mediaset, a private broadcaster in Italy, shows the migration toward digital TV is perceived only apparently as an evolution, but most effectively as a revolution. The migration of Mediaset to digital TV is characterised by two phases represented in Table 7.4.

Within the TV production systems, migration towards the digital occurred rather softly.

The first 15 years of operation were featured by **analog technologies,** which, introduced in 1978, are still present in some Mediaset sectors. The present production process is based upon them. In 1996 Mediaset began to transform its facilities to digital, as an evolutionary technology. This process had a strong technical and market relevance (analog equipment soon became

Table 7.4: Migration Toward Digital TV: Mediaset Experience

domains	EVOLUTION	REVOLUTION
Analog TV	1978	
Digital TV as Evolution	1996	
Digital TV as Innovation		1999
MEDIASET case	1° phase	2° phase

Source: Adapted from Mediaset, 2001

obsolete), but the production process was unaffected with respect to analog technology.

In 1999, on the heels of ICT, the digital innovation technology era was born. This implied for Mediaset a deep change of perspective and a reengineering of the production process which required new jobs and skills needs. The new approach changed the real life of a broadcaster in a digital TV perspective. The areas impacted by the migration process are summarised in Table 7.5.

Table 7.5: The Areas Impacted by Digital Migration

areas	FROM	TO
role in the supply chain	TV broadcaster	content provider / packager
market & business chances	mono-media operator	multi-media operator
engineering skills	TV equipment knowledge base	mixed TV tech.& ICT
competences	TV programs only	interactive programs & services
program creation	linear programs	non-linear contents
content management	TV programs oriented	asset oriented

Source: Adapted from Mediaset, 2001

For each area there are some main critical changes which have to be considered; these are summarised in Table 7.5.

a. *Role in the Supply Chain*
   From TV broadcaster to content provider/packager
   Contents are the core in the new digital TV perspective. Re-use of contents and re-packaging capability is a must for the new supply chain in order to provide new media services (Internet, broadband, mobile) and new digital TV services (Enhanced Digital Television, Interactive Digital Television, Multimedia Home Platform MHP Services, Multithematic PPV channels, etc.). For this reason an extended multimedia, audiovisual, archive-centric organisation should be preferable to standard TV program-centric broadcasting service.

b. *Market & Business Chances*
   From mono-media to multimedia operator
   In a TV program's production process, the audiovisual material is essential, but every single piece of information collected should be stored for a later re-use in a multimedia domain:
   - casting bibliography, photos, and info;
   - TV program scripts, stories, rough cuts, backstage footage, etc.;
   - charts, graphics, geo maps, location info, GPS positioning, etc.;
   - documents, speeches, voices, 3D virtual scenes, background sound effects, soundtracks, music, songs, etc.

   Re-use of contents, re-packaging capability, and the right management are musts for increasing revenues in the new business scenario.

c. *Engineering Skills Changes*
   From TV equipment knowledge base to mixed TV technology and ICT
   TV facilities designers have to move their knowledge base and competences to the ICT domains:
   - general purpose PC & server-based technology will replace dedicated, old-fashioned TV equipment (black boxes);
   - workflow analysis and production process re-engineering are essential;
   - competences in programming languages and OS;
   - software module development and GUI design: visual clearness, usability;
   - internetworking and network shared resource behaviours;

- role of old-fashioned TV equipment manufacturers changes; the new role of system integrators;
- IT and engineering departments: cooperation, collision, or convergence.

d. *Competences*
   From TV programs only to interactive TV programs and services
   Broadcasters know very well how to make exciting TV programs, either in analog or digital domains, but digital TV is much more:
   - *Interactive Programs*: Broadcasters, format inventors, and TV program creators have to face the creation of a new deal between their programs and the viewers. They have to find the way to involve them in a more appealing and emotional experience to play along with the TV program itself.
   - *Interactive Services*: Soon the Interactive Digital Television (iDTV) set and new generation set top box will be capable of offering digital advanced services to viewers. A new mass market perspective will come. Broadcasters, content, and service providers operators will play a new role, competing in this new market arena.

e. *Program Creation*
   From linear programs to non-linear contents
   In the next 10 years, linear programs will probably be detonated to become a prehistoric discovery for broadcasters. The old-fashioned way to create linear contents will come to be replaced by a non-linear interactive way to create interactive contents, thanks to the new generation of script editors and story makers, very well appreciated nowadays in the videogames industry. This non-linear content will be more ideal for future interactive TV programs and services.

f. *Content Management*
   From TV programs oriented to asset oriented
   An archive-centric organisation, which wants to play a significant role in a multimedia multidelivery publishing platform, either in the new media or in the digital TV market scenario, would preferably be focused on its media content management concept. There would be no more audiovisual tapes repository, as for ancient Videoteque, but a fully featured digital distributed archive integrated system, operated by a virtual centralised media asset management tool.

Old-fashioned tape labels, shelf, textual search, and retrieval mainframe-based legacy systems are going to be replaced soon by modern assets managers, proxy contents, metadata descriptors, indexers, etc.

Mediaset Innovation Projects

Mediaset has developed a number of innovative projects following the audio-video contents digitalisation process, as well as the creation of digital archives. Figure 7.10 highlights the value chain areas considered.

Figure 7.10: The Areas Impacted by Digital Migration

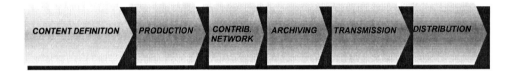

Digital TV Innovation Project in the Production Area

The introduction of digital technology seems to be particularly positive in news productions.

Digital TV Innovation Projects in the Production area include the fully featured Digital Newsroom for the Mediaset Headline News programs, where a journalist can look for pictures and set up his report directly from his desk. This project is made up of the following:

- Open Media, the newsroom system;
- Clipedit, the journalist digital video editor;
- more than 150 operational workstation;
- efficiency and cost-effective targets;
- new business oriented.

Through this project, the following targets were pursued by Mediaset:

- EFFICIENCY: through elimination of physical supports management;

- EFICACY: guaranteeing access to the same pictures by several users at the same time, with the chance of changing features also when close to broadcast time;
- NEW BUSINESSES: ability to use products made on any possible media (Themed channels, the Internet, etc.).

Open Media

Open Media is the management system of the newsroom, the tool through which a journalist receives all messages from information agencies, and writes his/her piece which is then read and authorised by the chief editor; who then finds its place in the TV broadcast.

Open Media is at present distributed on over 150 workstations: from editors' offices to line, production, and news studios. The journalist can receive information from press agencies, write his/her piece, and retrieve past information on the same topic. Management of both the editing process automation (workflow) as well as the news schedule can be achieved.

The Open Media system controls the on-air video servers to broadcast the news and is integrated with the news Digital Archive. The system produces five news bulletin editions per day.

Clip Edit

The journalist desktop editor is the tool that allows a journalist to work directly on his/her personal computer, as well as to see all the audiovisual contents shot by the ENG crew, acquired from video agencies, or broadcast from local stations. It also allows the journalist to select the most significant pictures, to set them and dub them for a given report.

An easy yet powerful graphic interface allows the journalist to carry out from his/her desk all the production phases which previously required him/her to move around in different premises. This includes:

- 15 filling workstations (video ingestion) (ENG crew shots, local, international agencies);
- 80 browsing, editing, and voiceover workstations;
- three craft editors for high-quality post-production of the audiovisual material;
- a daily server with 120 hours of high-quality video material online;
- a clip server for fast editing and voiceover of audiovisual material;
- two on-air servers for transmission.

Digital TV Innovation Project in the Contribution Area

RAV & FAV (video application networks—high-speed fiber) is a fiber-optic infrastructure interconnecting Mediaset production campuses (nodal points):

- *Cologno Monzese Center:* production center for entertainment programs' production;
- *Milano 2:* the news center;
- *Segrate:* broadcasting and Mediaset logistics center;
- *Roma Palatino* and soon also some local centers.

RAV & FAV is an high-speed optical network (Figure 7.11) with 100 optical fibers for the high-speed transport of all audiovisual contents among the network nodes (campus) and a powerful logic transport infrastructure based on the Gigabit Ethernet technology with a 4 Gigabit backbone. Quick and capillary widespread, physically separated from the corporate LAN but interconnected through specific gateways, it is widespread in all the buildings of the campuses.

Over 450 connections in "fiber desktop" technology are operational along the whole network which is bound to replace band physical transport in the future within the intermediate stages of production.

Figure 7.11: Optical Fiber Network

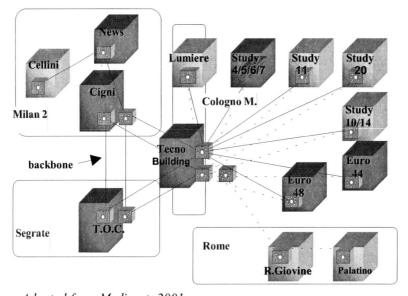

Source: Adapted from Mediaset, 2001

Digital TV Innovation Project in the Archiving Area

The Digital Archive is a strategic project for Mediaset. Audiovisual contents become, together with media data, an extremely important asset to feed all distribution links which are cropping up in the world of hypermedia (digital broadcasting TV, broadband Internet, mobile phone, UMTS, etc.) within the values featuring all the commercial offerings available: free TV, Pay TV, Pay-Per-View, On-Demand, etc.

D.a.i.s.y., digital archive integrated system, is a system of localized robotic libraries close to production—a large virtual, audiovisual, and multimedial archive made up of a number of robotised libraries physically detached. The archive has an automated and assisted classification of more than 400,000 hours of digitalised programs.

Management of this huge wealth of pictures and metadata takes place through a system of media asset management, allowing for the following:

- improve corporate exploitation of pictures through sharing;
- distribute on a real-time basis audiovisuals on all distribution platforms;
- allow for contents re-packaging.

The NEWS archive is the milestone of the Mediaset Digital Archive project. It includes up to 50,000 hours of audiovisual material broken down into three content quality levels:

- 30 Mbit/s for quality broadcasts;
- 1 Mbit/s to access and select materials (browsing);
- 150 kbit/s for New Media and its applications, and automated indexing of images and speech to text.

Over 5 terabytes of online material to file materials in transit from the robotised library and management of metadata database are also available.

In the future, all transport and handling procedures of contents among remote locations will be totally automated and integrated into the production process. The archive system is asset oriented, that is to gather all information connected with contents, i.e., video and metadata of the story, journalist's description, broadcasting schedule.

Digital TV Innovation Project in the On-Air Area

Digital broadcasts are a high-impact technological project. By January 2002 Mediaset was to have three digital on-air networks. Two power video servers controlled by two parallel automation systems will carry out the playout of the programs of the three networks according to a schedule developed and completely assembled by multimedial and audiovisual technologies, directly from the workstations of the assemblers of the broadcasting schedule.

A sophisticated emergency control system can detect any abnormality in each one of the four digital broadcasts and shift the whole schedule on the emergency broadcast.

Operators will work on totally redundant control systems (double), as happens in a jetliner cockpit.

The equivalent of over 15 days of programs are contained and stored into the two video servers for each network. A syncronisation system continuously checks that both servers have the same high-quality content, while a third video server hosts a low-quality copy for processing and schedule control purposes.

CONCLUSIONS

The findings of the present section of the study allow for an understanding of the advantages offered by the implementation of content management. There are many new opportunities and additional benefits created by the process of digitalization which can be summarized in the following:

- increase revenues and reduce costs;
- eliminate duplicate inventories of the same content;
- reduce shipping and storage costs;
- reduce the costs to reach customers;
- improve brand management;
- create and enforce consistent brand strategies;
- increase brand awareness by leveraging consistent content across the enterprise;
- reuse of content across marketing and sales channels reinforces and disseminates brand.

It's also important to note the improvement to processes in terms of:

- efficiencies and savings through value chain integration, reduction of duplicated effort, and the provision of consistent, accurate and current data;

- improvement of workflow management, control, visibility, reporting, and quality assurance;
- separation of content from format, allowing support for additional channels at very little incremental cost;
- improvement of communications and collaboration;
- effective interaction within and between organizations, vendors, suppliers, distributors, and customers;
- creation of an environment of increased creative collaboration, resulting in improved quality at lower cost.

Those companies that seize the opportunities to define new markets, channels, products, and revenue models, based on digital capabilities, will define the future shape of the industry. The leaders in the entertainment and media industry of the (near) future will be those companies that take digital capabilities in the core of their business.

Media and entertainment companies that do not seize the opportunities, or that dabble with digital capabilities in isolated sections of the business, will be followers. They will be unable to compete across digital channels with their existing brands and will fail to attract new brands because of this inability.

ENDNOTES

[1] Bowler, J. (2000). DTV content exploitation. What does it entail and where do I start? *New TV Strategies,* 2(7), 7.

[2] Pagani, M. (2001). Content management for a digital broadcaster. *Proceedings of Managing Information Technology in a Global Economy, 2001 IRMA Conference,* p. 1062.

[3] This theoretical model regards the firm as a set of activities, each of which produces value for the end user. Porter, M. (1985). *Competitive Advantage: Creating and Sustaining Superior Performance.* New York: The Free Press.

[4] Abell (1980) defines a business according to three dimensions: customer target group ("who" has to be reached); the functions offered to the customers ("what" customers want); technologies adopted ("in which way" customer needs can be satisfied). With reference to each of these three dimensions, the business is then defined according to the scope and differentiation of them. Abell, D. (1980). *Defining the Business: The Starting Point of Strategic Planning.* Englewood Cliffs, NJ: Prentice-Hall.

Chapter VIII

Digital Rights Management

INTRODUCTION

Digital Rights Management poses one of the greatest challenges for multimedia content providers and interactive media companies in the digital age in order to be able to monetorize their interactive products and services catalogs.

In a digital environment, digital contents (video, audio) can be distributed across different media (TV, radio, Internet). Organizations need to take into account some legal issues:

- control of copyright and rights management;
- regulation of content;
- regulation of access;
- customer data and consumer protection.

There are many problems about conflict between the territorial allocation of broadcast rights and the physical coverage of the signal. As illustrated in the previous chapters, satellite transmission has a much greater potential problem with audiovisual piracy than either terrestrial or cable due to the wide footprint of satellite broadcasts. In respect to out-of-area reception, cable networks offer the most controlled signal distribution since they comprise a closed environment. For terrestrial broadcasting the most typical issue is "overspill," when the same signals can be received in an adjoining territory. In highly cabled territories, the custom has developed of redistributing the overspill signal as widely as possible.[1] In these cases collecting societies will administer the copyright payments.

The Net poses challenges both for owners, creators, sellers, and for users of Intellectual Property, as it allows for essentially cost-less copying of content.

Digital files can be easily copied and transmitted, and today we already see serious breaches of copyright law. In an analog workflow DRM focused on security and encryption as a means of solving the issue of unauthorized copying, that is, lock the content and limit its distribution to only those who pay. This was the first generation of DRM, and it represented a substantial narrowing of the real and broader capabilities of DRM.

The second-generation of DRM covers the description, identification, trading, protection, monitoring, and tracking of all forms of rights uses over both tangible and intangible assets, including management of rights holders' relationships[2] (Iannella, 2002). It is important to note that DRM technologies enable secure management of digital processes and information. Digital Rights Management refers to the management of *all* rights, not only the rights applicable to permissions over digital content.

After giving a definition of Intellectual Property and Digital Rights Management, this chapter discusses the functional architecture domains which cover the high-level modules or components of the DRM system that together provide an end-to-end management of rights. Then the chapter provides a description of the typical Digital Rights Management value chain in terms of activities and players involved along the different phases examined.

The purpose of the chapter is to try to understand the impact of the adoption of Digital Rights Management on the Value Chain and to try to analyse the processes in the context of the business model. The chapter concludes with a discussion of the main benefits generated by the integration of Digital Rights Management and proposes the most interesting directions for future research.

INTELLECTUAL PROPERTY: A DEFINITION

Intellectual Property can be defined as information that derives its intrinsic value from creative ideas. It is also information with a commercial value.[3]

Intellectual Property Rights (IPRs) are bestowed on owners of ideas, inventions, and creative expressions that have the status of property.[4] Just like tangible property, IPRs give owners the right to exclude others from access to or use of their property.

The first international treaties covering Intellectual Property Rights were created in the 1880s and are administered by the World Intellectual Property Organisation (WIPO), established in 1967. These treaties are:

- The Paris Convention for the Protection of Industrial Property
- Berne Convention for the Protection of Literary and Artistic Works

The newly revealed physics of information transfer on the Net has changed the economics and ultimately the laws governing the creation and dissemination of Intellectual Property. The Net poses challenges for owners, creators, sellers, and users of Intellectual Property, as it allows for essentially cost-less copying of content. The development of the Internet dramatically changes the economics of content, and content providers operate in an increasingly competitive marketplace where much content is distributed free (see the Napster phenomenon).

KEY INTELLECTUAL PROPERTY RIGHTS ISSUES

There are many issues that organisations need to address to fully realise the potential in their intellectual property. They are summarised in the following:

- *Ownership:* clear definition of who owns the specific rights and under what circumstances;
- *Distribution*: definition of the distribution strategy focused on a small trusted group or the mass market;
- *Protection*: definition of the content the organisations need to protect and the level of protection required;
- *Globalisation*: because of the slow harmonisation across countries regarding protection of Intellectual Property Rights, what is acceptable in one country may have legal implications in another;
- *Standards*: understanding what standards are in development and how these may affect system development.

Ownership in the Digital World

The first critical issue that organisations need to address is to protect their ownership in the digital world.

Prior to the Internet and Digital Broadcasting, it was difficult to dissect Intellectual Property (the message) from the medium in which it travelled. The development of the Internet has been the most impressive revolution in the new society of information.

Many regard the Internet as a threat to Intellectual Property and seek to change laws to protect their interests, yet this new medium has brought about new business opportunities for those involved with Intellectual Property.

Rights owners fear a loss of control of their core content: in the case that the same content is distributed free, there is a consequent loss of the intrinsic value of rights (see the Napster phenomenon). Other threats are represented by the cannibalisation speed, which lowers video market value, and other limits on rights as different types of content have different ownership issues:

- film rights are often acquired on a world-wide basis;
- publishing rights are likely to be territory specific;
- textual material may well be associated with well-known illustrations which are under their own copyright;
- music is usually owned by the author or record companies.

To date, precious few of the above threats have an inkling of an answer, and organisations need new copyright laws.

Distribution

In a digital workflow, any piece of electronically represented Intellectual Property can be almost instantly anywhere in the world. Controlling copies becomes a complex challenge.

There are two main options: 1) put in place very tight control measures, limiting distribution to a small, trusted group or 2) have little or no control measures in place, but rest assured that eventually the "product" will find its way to a large non-paying audience.

Digital broadcasting offers the opportunities to exploit assets over multiple channels, but only if the rights can be safeguarded and monetorized. Considering the UK case study, regulation today[5] affects:

1. *Web content alone:* In the UK, broadcasters cannot advertise or trail websites or other interactive content on air, unless "program support" is largely non-commercial.
2. *Overlays on TV shows* are regulated only if inside the broadcast stream; they are unregulated if outside the broadcast stream to the same screen (i.e., Blue Square on Onnet or betting on Sky via Sky Text).
3. *Web functions on TV* are regulated[6] only if they are broadcast on TV platforms (primary concern is "enhanced TV"), but there is no regulation if they are delivered by another mechanism. Some tricky areas refer to:
   - separating ads and editorial;
   - kids programming;
   - sponsored news.
4. *TV functions on the Web:* Any of the above delivered via the Web is unregulated other than the on-air trailing of such content.
5. *PVR functionality* is not regulated.

Protection of Intellectual Property

This critical issue refers to the definition of the content the organisations need to protect and the level of protection required.

A myth exists that, if work is published on the Internet, it is within public domain, but this is not true. The law has changed and it now states that neither publication nor a notice of any kind is required to protect works today.

Creators and content providers must decide what price is worth paying to protect their Intellectual Property (e.g., it is not worth spending $5 to protect something worth $2). There is still a feeling that, as no protection is 100% foolproof, why spend money on something that can't be guaranteed.

For low value content most owners will allow content to be available unprotected to encourage mass distribution and copying. Revenue will be realised through subsidiary services or products.

Protected content will be higher quality versions of content that is available free of charge in more primitive formats.

Globalisation

Harmonisation across countries regarding protection of Intellectual Property Rights remains low. If a product is launched on the Internet, then it is available around the world. This can cause problems if the distributor only has rights to the content in a particular region.

Legislation may require that a particular work is not shown in certain countries. In this case rights ownership can be problematic particularly if the work is distributed globally. Enterprises need to control the rights for the appropriate uses and territories.

There is also the issue of cultural differences between countries. Many European countries recognise all the rights of Intellectual Property that we take for granted; but in China and in other Asian countries, copyright is grafted onto a culture that has valued initiation as the foundation of learning and of the transmission of culture across generation.

Standards

There are a lot of technical and industry standards: MPEG2, MPEG4, REAL, Windows Media, XrML, ODRM, XMCL, MP3, PDF, EBX, OEB, etc. Standards are emerging at different paces according to the maturity of the medium: video, publishing, or music. Further, standards for format are blurring with standards for right protection.

Technology companies are hedging their bets. Having been stung in the past over format wars, many now wait for the content owners to make the call. There are a number of different organisations and committees trying to come up with the standard which they hope will be adopted by the majority. Firms that set important standard, create the dominant designs, or control a dominant architecture may be even more important in converging industries then they were in PCs or consumer electronics.[7]

DIGITAL RIGHTS MANAGEMENT: FUNCTIONAL ARCHITECTURE

Digital Rights Management (DRM) covers the description, identification, trading, protection, monitoring, and tracking of all forms of rights uses (Figure 8.1) over both tangible and intangible assets, including management of rights holders relationships.

At the heart of any DRM technology is the notion of a rights model. Rights models are schemes for specifying rights to a piece of content that a user can obtain in return for some consideration, such as registering, payment, or allowing her usage to be tracked.

Digital Rights Management (DRM) can be defined as the secure exchange of Intellectual Property, such as copyright-protected music, video, or text, in digital form over channels such as the Web, digital television, interactive TV, digital radio, and the much-talked-about 3G (third-generation) mobile or other electronic media, such as CDs and removable disks. The technology protects content against unauthorised access, monitors the use of content, or enforces restrictions on what users can do with content.[8]

Digital Rights Management (DRM) allows organisations that own or distribute content to manage the rights to their valuable Intellectual Property and package it securely as protected products for digital distribution to a potentially paying, global audience.

DRM technologies provide the basic infrastructure necessary for protecting and managing digital media, enterprise-trusted computing, and next-generation distributed computing platforms, and they allow content owners to distribute digital products quickly, safely, and securely to authorised recipients.

On the basis of the results discussed in Chapter 7, the digital Media Management Value Chain can be described by the following activities: acquisition, media asset management (encode, manage and archive assets, manage workflow), delivery management (see Figure 8.2).

Figure 8.1: Fundamental Types of Rights

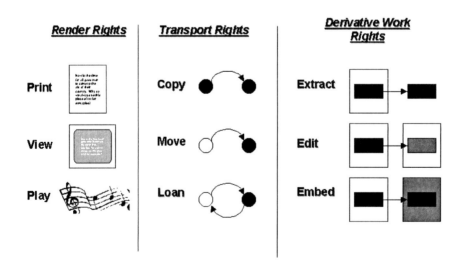

Figure 8.2: Typical Digital Media Management Value Chain Activities

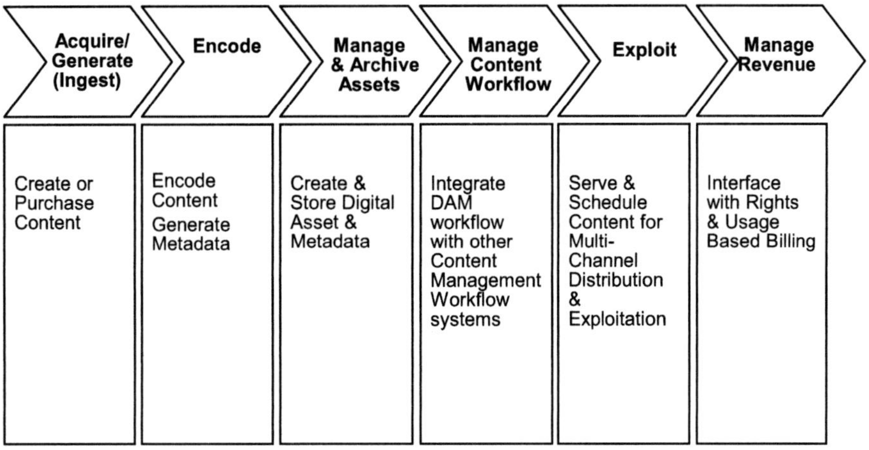

Acquire/Generate (Ingest)	Encode	Manage & Archive Assets	Manage Content Workflow	Exploit	Manage Revenue
Create or Purchase Content	Encode Content Generate Metadata	Create & Store Digital Asset & Metadata	Integrate DAM workflow with other Content Management Workflow systems	Serve & Schedule Content for Multi-Channel Distribution & Exploitation	Interface with Rights & Usage Based Billing

Each of these areas plays a key role in building digital rights-enabled systems:

1. *Intellectual Property (IP) Asset Creation and Acquisition*: This area manages the creation and acquisition of content so it can be easily traded. This includes asserting rights when content is first created (or reused and extended with appropriate rights to do so) by various content creators/providers. This area supports:
   - **rights validation:** to ensure that content being created from existing content includes the rights to do so;
   - **rights creation:** to allow rights to be assigned to new content, such as specifying the rights owners and allowable usage permissions;
   - **rights workflow:** to allow for content to be processed through a series of workflow steps for review and/or approval of rights (and content).

2. *Intellectual Property Media Asset Management*: After the finished content is bought, this area manages and enables the trade of content. The digitalisation of the television signal and the storage of the materials that have been purchased need to manage the descriptive metadata and rights metadata (e.g., parties, uses, payments, etc.). This supports:

- **repository functions:** to enable the access/retrieval of content in potentially distributed databases and the access/retrieval of metadata;
- **trading functions:** to enable the assignment of licenses to parties who have traded agreements for rights over content, including payments from licensees to rights holders (e.g., royalty payments).

3. *Intellectual Property Asset Delivery Management*: This area manages the distribution and usage of content through different platforms (TV, radio, Web, 3G mobile) once it has been traded. This includes supporting constraints over traded content in specific desktop systems/software. This area supports:
   - **permissions management:** to enable the usage environment to honor the rights associated with the content. For example, if the user only has the right to view the document, then printing will not be allowed;
   - **tracking management:** to enable the monitoring of the usage of content where such tracking is part of the agreed-to license conditions (e.g., the user has a license to play a video 10 times).

Figure 8.3: DRM Functional Architecture

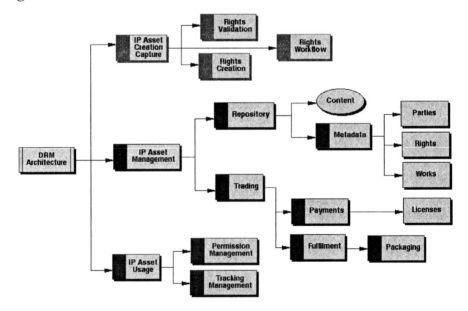

Source: Iannella, 2002

The broad areas required for DRM are characterised by different functions which can be expressed by the Functional Architecture (Iannella, 2002) that provides the framework for the modules to implement DRM functionality (see Figure 8.3).

DIGITAL RIGHTS MANAGEMENT: VALUE CHAIN ACTIVITIES

After defining the functional architecture, we can describe at a first level the DRM Value Chain, identifying the activities supported by each segment and the players involved. At a second level we try to understand the DRM processes in the context of the Digital Content Management Value Chain and the impact of these processes on the business model.

The Digital Rights Management Value Chain can be described by six main segments: contract & rights management, rights information storage, license management, persistent content protection, clearing house services, billing services (cfr., Figure 8.4).

Each segment along the Value Chain is characterised by specific activities:

1. *Contract and Rights Management*: the registration of contract terms and rights, tracking of usage, and payment of royalties (residuals).
2. *Rights Information Storage*: the storage of rights information (e.g., play track five times) and usage rules (e.g., if they have a UK domain and have paid) as well as rights segmentation and pricing structures.
3. *License Management*: the management and issuing of licenses in line with the Rights and Conditions. Without a license the consumer can't use the content.
4. *Persistent Content Protection*: the use of encryption, keys, and digital watermarking to securely package digital content. This stops the consumer from using the content without a license.

Figure 8.4: Digital Rights Management Value Chain Activities

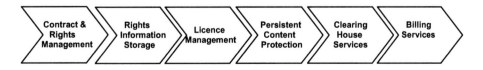

5. *Clearing House Services*: managing and tracking the distribution of the packaged content and the license in line with the defined Rights Information.
6. *Billing Services*: charging consumers for purchased content and payment to parties within the Value Chain.

Various players in the market are affected by Digital Rights Management, in particular all those operators involved with ideation, creation, realization, enabling, and handling of contents. The role of these operators is to provide contents successively broadcast on network channels, radio, the Internet, and other media. The different players involved along the Value Chain can be summarised in the following (see Figure 8.5):

- *Content Author/Creator* (e.g., Canal+, Reuters) or any other media or non-media company that produces content for internal or external use;
- *Content Publisher/Aggregator* (e.g., IPC Magazines, Flextech Television, BskyB) who buy various content and aggregate it into channels aimed at a particular lifestyle or niche;
- *Content Distributor*, e.g.. W.H. Smith News, Nokia, BskyB, NordeaTV;
- *Service Provider/eTailer*, e.g., T-Online.com, Yahoo.com.

In the first phase of the Value Chain, *Contract & Rights Management*, after the author has created the content, the aggregator packages it in a container which provides persistent protection, enforcing the rights which the author has granted. This may be written as an applet that travels with the content which will be encrypted.

In the second phase, *Rights Information Storage,* the aggregator specifies the rights which apply to the content using products such as IBM, EMMS,

Figure 8.5: Digital Rights Management Value Chain Activities and Players

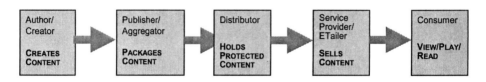

or Microsoft which encode the rights using an XML-based standard such as XrML or XMCL.

In the third phase, *License Management,* the consumer purchases the rights to use the content, the eTailer obtains the content from the distributor and requests a license from the content clearing house. The license may well be written in XML and may travel with the packaged content or separately.

The fourth phase is *Digital Asset Protection.* The consumer cannot access the content without the license. The Media Player, e.g., Real Video Player, interprets the license and enforces the rights granted to the consumer. That may include how many times or for how long they can access the asset, whether they can duplicate it or "rip" CDs from it.

With reference to the Digital Content Management Value Chain described above, DRM processes play an important role in the encode activity (Contract Management & Rights Storage), management and archive assets (Encrypt/Package Content Asset Protection), management workflow (Key & License Generation), and exploit (Key/License Management & Clearing House Services) (Figure 8.6).

At a second level of analysis, we try to understand Digital Rights Management processes, described above, in the context of a business model.

Figure 8.6: DCM Value Chain Activities & DRM Processes

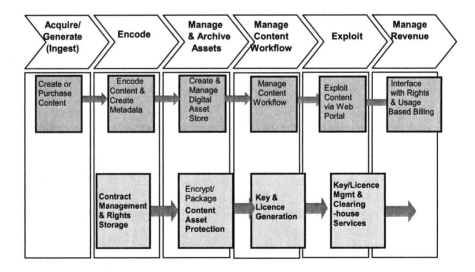

Figure 8.7: The Process in the Context of the Business Model

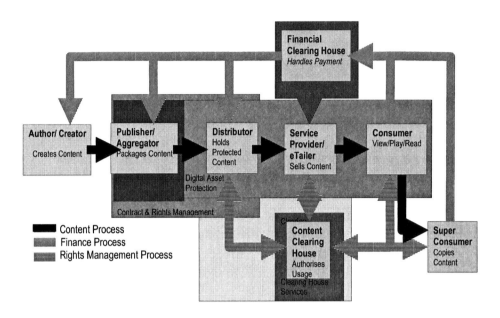

All the processes analysed can be grouped in three broad categories (see Figure 8.7):

1. **content processes**: include all systems and processes that enable the creation/capture, formatting, packaging, protection and storage of trusted channel content, control of when and how content is released, where and when it gets used, and how long it remains there;
2. **finance processes**: include all systems and processes (payments) between the Financial Clearing House and the other players along the Value Chain (author/creator, publisher, distributor, consumer);
3. **Rights Management processes**: include all Content Clearing House services which authorise the usage of rights to distributors, service providers, and consumers.

DIGITAL RIGHTS MANAGEMENT BENEFITS

There are many benefits generated by Digital Rights Management. The first main area concerns all benefits related to Contract & Rights Management.

Digital Rights Management allows a better management of bought-in content rights to maximise return on investment, and it avoids potentially expensive misuse. A better management of content is also allowed by means of the creation of what-if scenarios for potential new revenue streams on the basis of the cost and ownership of content rights.[9] New contents can no longer be constrained to budget on a cost basis, but they can now budget on the basis of forecast rights revenue.

Digital Rights Management generates benefits also with reference to the sale of rights, allowing a retained control of the sale of content rights and the conditions of sale in mass or niche distribution environments. DRM also allows the establishment of flexible business models for digital content sales: rental or purchase, play or edit, burn to CD, or just play online.

Figure 8.8: DRM Market Development

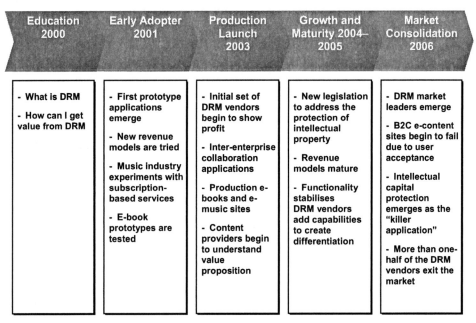

Source: Gartner Research. (2001). *Content Management Providers: Timetable Toward DRM*, July.

Segmentation of content allows the creation of different versions with different rights and conditions for different markets. Information that has previously been stored in a vast number of separate databases can now be merged, sorted, and analysed, resulting in the creation of a personal profile or data image of a subject based on his or her electronic data composite.[10]

The next three to five years will be critical for the Digital Rights Management market as it matures and becomes profitable (Figure 8.8).

Consumer acceptance will determine the success of the market. The critical factor for vendors is to address the right market (publishing, audio, video, and software).

CONCLUSIONS

Digital Rights Management allows organisations to manage the rights to their valuable Intellectual Property and package it securely as protected products for digital distribution. It's not solely about technology; Digital Rights Management works across the people, process, and technology boundaries. The key issues include:

- handling of complex sets of rights within each asset;
- rights licensing and management, and digital rights protection;
- understanding and design of revenue generation and collection models;
- standards—flexible rights languages and content formats;
- globalisation—territorial issues, both legal and commercial;
- ownership of rights.

Digital Rights Management is emerging as a formidable new challenge, and it is essential for DRM systems to provide interoperable services. Solutions to DRM challenges will enable untold amounts of new content to be made available in safe, open, and trusted environments. Industry and users are now demanding that standards be developed to allow interoperability so as not to force content owners and managers to encode their works in proprietary formats or systems.

The market, technology, and standards are still maturing. Digital Rights Management should be considered an integral part of a company's Digital Media Management framework.

ENDNOTES

[1] CDG Eurostudy. (1999). *Technical Perspective.* London, England: *CDG Report*, p. 46.

[2] Iannella, R. (2001). Digital Rights Management (DRM) architectures. *D-Lib Magazine*, 7(6).

[3] See Campobasso, G.F. (1991). *Diritto Commerciale.* Turin, Italy: UTET, p. 194.

[4] Scalfi, G. (1986). *Manuale di Diritto Privato.* Turin, Italy: UTET, p. 61.

[5] See also Enser, J. (2001). Legal issues in the use of content. Olswang, Speech at *Content for Digital Devices Conference*, London, September 25.

[6] ITC Paper, February 2001.

[7] Anderson, P. & Tushman, M.L. (1990). Technological discontinuities and dominant designs: A cyclical model of technological change. *Administrative Science Quarterly,* 35: 604-633; Ferguson, C. & Morris, C. (1993). *Computer Wars: How the West Can Win in a Post-IBM World.* New York: Random House.

[8] Forrester, 2002.

[9] See also Burke, D.C. Digital Rights Management: From zero to hero? Cap Gemini Ernst & Young, Speech at *IBC Nordic Euroforum Conference*, Stockholm, Sweden, February 20.

[10] Tapscott, D. (1999). Privacy in the digital economy. In Tapscott, D. (Ed.), *The Digital Economy.* New York: McGraw Hill, p. 275.

Chapter IX

Conclusions

Interactive multimedia and the so-called information highway, and its exemplar the Internet, are enabling a new economy based on the networking of human intelligence. In the digital frontier of this economy, the players, dynamics, rules, and requirements for survival and success are all changing.

The difficulties in sustaining the business models, which have been recently created worldwide, makes this topic extremely relevant in order to understand the sustainability of competitive advantage in the television environment.

How will the market for digital interactive television develop?

We are going to be consuming more communication, both broadcast and narrowcast, and at least for the immediate future this communication will take digital forms. Costs and prices are falling because of technological progress in processing and transmission, and because of increased supplies of spectrum from the government, not merely for economies of scale in sharing pipelines.

Conventional television seems to satisfy a demand for which interactive television is not a very good substitute. Many studies in the economic literature of leisure time use (Robinson & Goeffrey Godbey, 1997) shed light on the demand issue, and they affirm that part of the allure of television is freedom of choice and interactive television may actually be less appealing to people if they must invest more energy and imagination.

Managers must not forget that the final player of the iTV value chain is obviously the end user, whose behaviour and preferences are critical factors determining the success of the other players and of the whole industry.

Future demand and penetration of interactive TV is expected to grow very fast. Forecasts assert that Europe's iTV penetration will reach 44% of European households by 2007, up from only 11% in 2002,[1] with four countries (UK, France, Spain, and Italy) driving the growth and accounting for 70% of Europe's iTV households. When considering these projections it is useful to remember the "crossing the chasm"[2] paradigm in the technology-adoption lifecycle model. Crossing the chasm is jumping that empty area between the innovators' segment and the early majority. The early adopters are iTV enthusiasts and are always looking forward to experience technology innovations. Being the first, they are also prepared to bear with the inevitable bugs that accompany any innovation just coming to market. By contrast, the early majority is looking to minimise the discontinuity with the old ways, and they do not want to debug somebody else's product. By the time they adopt it, they want it to work properly and to integrate appropriately with their existing technology base. Visionary early adopters and the pragmatist early majority have completely different frames of mind about technology; because of these incompatibilities, early adopter surveys don't help to really understand and to predict accurately how consumer behaviour might change as a response to the introduction of the new technology.

WHAT CAN WE EXPECT FROM INTERACTIVE DIGITAL TV?

As with all things new, until the consumer has first-hand experience of a medium or product, take-up can only ever be uncertain. Even with this volume of users, content providers and brand marketers are wondering how they can:

a. capitalise on this sizeable audience;
b. make use of the interactive medium to reach out to and communicate with this audience;
c. make money online while not cannibalising their core business.

A multi-channel environment can further exacerbate increasing audience fragmentation, potentially vanquishing the "reach is all" mentality that has dominated the TV advertising business for years.

Digital TV is an ongoing reality, and for this reason, building a business strategy which accommodates change is vital.

Interactive TV services are providing welcome opportunities for brand marketers keen to pursue closer relationships with a more targeted audience, and with the promise of a new direct sale channel complete with transactional functionality. For broadcasters, garnering marketer support and partners can be a crucial means of reducing costs, providing added bite to marketing digital TV to consumers, while establishing new sources of revenue (based on carriage fees from advertisers, revenue shares for transactions coordinated via the digital TV platform, and payment for leads generation and data accrued through direct marketing).

At this stage in the interactive television market, there are more questions to be answered.[3] The issues at hand are:

- What are the prospects for interactive TV as a viable medium and (theoretically) in what kind of timeframe?
- Can interactive TV go any way to alleviating the problem of audience fragmentation? How will the traditional TV advertising model change (will advertisers demand a rate-card based on click-throughs, as with the Web)?
- How can interactivity open up new marketing opportunities (e.g., direct sales channels, direct advertiser-customer marketing relationships, discount-led marketing, leads-generation, data capture)?
- Who will take lead in developing content for interactive TV, both in terms of production and creative execution? Will it be broadcasters, who can exact greater control over the environment they actually built? Or will this be agency led?
- Who potentially has the best expertise to develop interactive advertising? Is it the traditional advertising agency, or the interactive specialist agencies, up to now more accustomed to building for the Web? Or will a new form of specialist agency emerge which combines all world views? Indeed, might advertisers themselves be able to more directly control these aspects and set up their own production units?
- To what extent will the non-interoperability of digital TV platforms actually act as a barrier to advertisers embarking on interactive advertising campaigns?

To date, precious few of the above questions have an inkling of an answer. The number of launch failures for interactive multimedia services in U.S. and

Europe up to now provides an example of the risk rate that is a feature of the new communication markets.

From a strategic point of view, the main concern for broadcasters and advertisers will be how to incorporate the potential for interactivity, maximising revenue opportunities and avoiding the pitfalls (many unknown) that a brand new medium will afford. It is impossible to offer solutions, merely educated guesses for how interactive TV will develop.

Success will depend upon people's interests for differentiated interactive services. Nevertheless, companies will need to be adept in a number of areas if they are going to challenge, and carve a role alongside, the strong digital TV players.

1. Firstly, the development of a clear consumer proposition is crucial in a potential confusing and crowded marketplace.
2. Secondly, the provision of engaging, or even unique, content will continue to be of prime importance.
3. Thirdly, the ability to strike the right kind of alliances is a necessity in a climate that is spawning mergers and partnerships. Those who have developed a coherent strategy for partnering with key companies that can give them distribution and content will naturally be better placed.
4. Finally, marketing the service and making it attractive to the consumer will require considerable attention, not to mention investment.

Next it is apparent, as we have already found on the Internet, that understanding consumer behaviour in interactive "channels" is crucial for fashioning interactive services. Knowing not only who is interacting but also when, where, and in what way they engaged, is invaluable. All of this, of course, is directly linked to determining what is the best way to get consumers to respond to the ad and interact with the brand.

The newness of the services, above all in the area of interactive services and the various forms of Pay-Per-View, require an intense (and costly) effort of persuasion directed at potential users.

The (monetary) costs terminal equipment and the (psychological) costs to get used to the new services seem to place a premium on perceived advantages, above all in countries where there is an abundant offering of free television.

The viewer does not invest in quantitative expansion so much as in qualitative expansion and improvement. This means that the viewer is prepared to spend money in order to access a variety of increasing alternatives, and to make a "personalised" choice within this variety.

Interactive service providers therefore have to create new markets for interactive television services, and above all create a critical mass of first users in order to be able to consolidate their market position.

Operators in the market are increasingly exploiting the possibilities offered by the new platforms (the Internet in particular) for pushing their own activities beyond the boundaries (geographical and productive) of the markets in which they have traditionally operated.

The flexibility of digital information makes it possible not only to offer more and richer traditional services (for example, digital radio and television or better quality mobile communications), but also a wide range of new services and applications. These run from electronic papers to online markets and catalogs, and from home banking to the use of multimedia websites as an internal means of communication or as a work tool.

In sum, the development of the market generated by technological innovations forces the individual television firm increasingly to know its positioning and the state of the dynamic competition of the moment. This book helps to understand the fundamental issues of interactive multimedia digital television in the new multimedia convergence, explaining the nature of convergence of various technologies and the changing business models that accompany this shift. Digital television is described and analysed as a clear example of such convergence.

This book helps the reader to understand how no economic analysis can leave out of account the relationships between technological innovation, dynamic competition, and the firm's market policy in the process of convergence that is currently going on in the television sector during the digital era.

ENDNOTES

[1] *Europe's iDTV Penetration Closes in on the PC Net.* Forrester Research, April 30, 2002.

[2] Moore, G. (1990). *Marketing and Selling High-Tech Products to Mainstream Customers*, London, UK: Harper Collins Publishers.

[3] Stewart, C. (1999). Minefield or goldmine: What can we expect from interactive TV? *New TV Strategies,* 1(1), 10-14.

Appendix

Acronyms and Abbreviations

ADSL	Asymmetric Digital Subscriber Line
API	Application Programming Interface
ATSC	Advanced Television Systems Committee (the American digital TV format)
ATM	Asynchronous Transfer Mode
ATVEF	Advanced Television Enhancement Forum
CTS	Cable Test System
DAVIC	Digital Audio-Visual Council
DOCSIS	Data-Over-Cable Service Interface Specification
DSB	Direct Satellite Broadcast
DSL	Digital Subscriber Line
DSM-CC	Digital Storage Media – Command and Control
DTT	Digital Terrestrial Television
DTV	Digital TV
DVB	Digital Video Broadcasting (-C: cable, -S: satellite, -T: terrestrial)

EPG	Electronic Program Guide
FSN	Full Service Network
FTTC	Fiber-to-the-Curb
HBO	Home Box Office
HCT	Home Computer Terminal
HDTV	High Definition Television
IA	Interactive
ISDB	Integrated Services Digital Broadcast (the Japanese digital TV format)
ISDN	Integrated Services Digital Network
iTV	Interactive Television
LMDS	Local Multichannel Distribution Service
MHEG	Multimedia and Hypermedia Experts Group
MOD	Movies-On-Demand
MMDS	Multichannel Multipoint Distribution System
MPEG	Moving Pictures Experts Group
NC	Network Computer
NVOD	Near-Video-On-Demand
PDA	Personal Digital Assistant
POTN	Plain Old Telephony Network
POTV	Plain Old Television
PPV	Pay-Per-View
PVR	Personal Video Recorder
SGI	Silicon Graphics

SMDS	Switched Multimegabit Data Service
SOD	Services-On-Demand
TBA	To Be Announced
VCR	Video Cassette Recorder
VOD	Video-On-Demand
YCTV	Your Choice Television

Glossary

Ad Views (banners)	Please also refer to impressions.
Ad Click Rate	This is the percentage of ad views (banners and datagems) that resulted in an ad click.
Ad Click	The number of times a banner is clicked on by a viewer.
ADSL	Asymmetric Digital Subscriber Line—uses existing copper wire telephone lines to deliver a broadband service to homes. It is one of the most viable forms of Digital Subscriber Lines due to its effectiveness over distance, i.e., it does not require the user to be as close to an exchange as other forms of DSL. Asymmetric refers to the fact that it provides a faster downstream (towards the consumer) than upstream (towards the exchange) connection. ADSL is "always on" and is considered to be the main rival to cable in Europe. At present, download speed is 516k downstream, but in the next two years this should increase to 4mb. *See* DSL.
Analog Data	Data represented by physical quantity that is continuously variable and proportional to the data
ARPAnet	Predecessor to the Internet. Developed by Defence Department in 1969.
Aspect Ratio	Term used to describe the width-to-height ratio of the television picture. Wide-screen TV uses an aspect ratio of 16:9, compared with the traditional TV aspect ratio of 4:3.

ATM	Asynchronous Transfer Mode—An advanced data transmission and switching protocol that greatly increases the capacity of transmission paths, both wired and wireless. ATM uses packets of fixed size and establishes "virtual" circuit connections.
ATSC	The Advanced Television Systems Committee is an international organisation, comprising 200 members, that is tasked with establishing voluntary technical standards for the next-generation television systems. ATSC Digital TV Standards include digital High Definition Television (HDTV), Standard Definition Television (SDTV), data broadcasting, multichannel surround-sound audio, and satellite direct-to-home broadcasting.
ATV	Advanced Television—An FCC term to designate what is now called DTV or digital television. *See* HDTV.
ATVEF	The Advanced Television Enhancement Forum—This is an alliance located in the U.S. consisting of companies representing all components of the television broadcast industry. Its basis is to move toward standardisation within the industry and to ensure Web protocols are included in standards set for iTV creation.
Audience flow	Term used when a viewer moves from one channel to another when the program changes, compared to those that remain with the original channel.
Backbone	That portion of communication network, such as the Internet, made up of very high-capacity trunks connecting switches or routers.
Back Channel	The "back channel" can also be referred to as a "return path" and describes when digital information is sent from the user's set top box via a telephone line to the broadcaster. This is usually facilitated via a dial-up modem, similar to the device found in the PC at home used for connecting to the Internet. This allows for two-way interaction between the broadcaster, and makes services like interactive game shows possible and also allows advertisers to monitor the areas the user has visited within their interactive campaign. Some platform providers, for example ONdigital, don't currently use a dial-up modem for the back channel and at present rely on the user to pick up the phone in the conventional sense. This means that

if an advertiser wants to build a product offer into their interactive campaign, then the user has to telephone through to a call centre to ask for the product.

Bandwidth   This concerns how quickly digital information can be passed along a network. The higher the capacity of the network, the faster it can travel. Interactive TV is delivered along one of the following networks: cable, satellite, terrestrial, or ADSL. Bandwidth is particularly important when you are trying to send or receive large amounts of information, and is particularly relevant when you are sending or receiving "bandwidth-hungry" information like video and audio.

Banner   A banner is a simple graphical device no larger than 346x46 pixels, which usually offers viewers the chance to "click-through" to see information on the advertiser. Banners are used predominantly on the Web and also within the walled garden of the iDTV platforms.

BARB   This is the key source of TV audience data in the UK. BARB is responsible for quantitative audience measurement and qualitative audience reaction or the audience's appreciation of programming. Viewing estimates are obtained from panels of television-owning households representing the viewing behavior of the 23+ million households within the UK. The panels are selected to be representative of each ITV and BBC region, and collectively provide a network sample of 4,485 households.

Bit   A binary unit of information or data derived from a choice between two equally probable alternatives, such as zero or one, on or off.

Bitstream   A sequence of bits transmitted on a communication channel.

Broadband   (1) A high-capacity communication link, wired or wireless, capable of transmitting the equivalent of multiple TV signals. *See* Narrowband. (2) Any communication channel or medium capable of data rates in excess of what can be achieved with a telephone line and an analog modem.

Broadcast   One to many communications, print or electronic. *See* Multicast; Narrowcast.

Broadcasting   As used herein, radio or television (video) transmissions.

Browser	A client program (software) on a computer, set top box, or other device that is used to look at various kinds information, including Web pages and interactive television interfaces. An example of a web includes Netscape and Internet Explorer. Interactive TV set top boxes use their proprietary browsers; e.g., DTV navigator used on CWC and Telewest.
Buffer	A mechanism for storing data temporarily because they are arriving faster than they can be processed.
Byte	A defined number of bits, usually eight, often corresponding to a letter or symbol, upon which computer operations are performed.
Cable Modem	A "high-powered" modem that permits one-way or two-way high-speed data communication over a cable television system for purposes such as Internet access at speeds of around 1.5 Mbps. Download rate is 27 Mbps. It does not need to dial-up like conventional modems and is therefore "always on," otherwise know as "impulse response."
Cache	A temporary store of data intended for use or reuse; for example, recently viewed Web pages that might be revisited.
C-Band	A portion of the electromagnetic spectrum designated by the FCC for, among other things, the first commercial satellite communications.
CD	A (digital) compact disc, originally for music; also used for computer data, in which case it is called a CD-ROM, for "read-only memory."
Cellular System	A wireless communication system in which relatively low power of focused transmitters reuse frequencies in non-contiguous geographic areas (cells).
Chat	Instant text communication over an electronic network between users, either anonymously or with known correspondents. Examples of chat include bulletin boards, chat rooms, and "instant messaging."
Churn	A term that describes the rate at which a Pay-TV service loses customers, typically represented as a percentage and measured on an annualised basis, i.e., 30% churn rate denotes a loss of 30% of the customer base yearly.

Circuit Switching	A communication network in which users are connected, through switches, using a channel dedicated to that use for the duration of the communication. A telephone system is an example of such a network. Packet networks, in contrast, are "connectionless."
Clarke Orbit	The orbit at an altitude of 22,300 miles above the equator at which a satellite is stationary relative to the Earth.
Click-Through	This is referred to when the user clicks on a banner advertisement within the walled garden or the broadcast stream and hyperlinks to a micro site. Advertisers are concerned with click-through rates, but this is only one of the ways of measuring the effectiveness of advertising.
Client	A computer or user in communication with a server.
Coaxial Cable	A broadband transmission line consisting of two cylindrical copper conductors arranged concentrically, separated by insulation.
Codec	Coder-decoder or compressor-decompressor. Hardware or software that serves as an intermediary between a computer and a digital transmission medium.
Commercial Impressions	The total audience for all adverts in a schedule.
Compression	Reduction of the bandwidth or number of bits needed to encode information, most commonly by eliminating redundant bits. A means of saving transmission time and storage space.
Concurrency	This is a condition which affects digital cable users only, which refers to the speed of the service received by users and is dependent on how many people are connected at any one time. The more subscribers that are connected to the exchange, the slower the service becomes.
Conditional Access	This refers to the manner in which channels are encoded or "scrambled." By controlling the operation of the unscrambling system via a prepaid access card or transmitted code, the broadcaster is able to control access to particular channels or services. Conditional access is typically used for Pay-Per-View and parental control.

DAL	Dedicated Advertiser Location (also described as a micro site)—A small site, usually consisting of a few pages or screens, which has a pre-determined lifespan. The purpose of the site usually supports a product launch or other similarly focused marketing activity.
Data	A collection of bits. The quantities or symbols on which computer and communication equipment operate, typically stored or transmitted in the form of electromagnetic energy. *See* Bit; Information.
DBS	Direct broadcast satellite television service, such as DirecTV or Primestar. *See* GEO.
Digital	A function that operates in discrete steps, such as "on" and "off." Because the physical world is continuous, such representations are approximations. Digital communications uses discontinuous, discrete electrical, optical, or electromagnetic signals that change in frequency, polarity, or amplitude.
DirecTV	A DBS service operated by General Motors' Hughes business unit. Shares a satellite with USSB; the combined system is called DSS.
Download	The process of retrieving data from a distant database; also, the data so retrieved.
DRAM	A type of random access memory chip.
DSL	Digital Subscriber Line—An ordinary telephone line improved by expensive equipment, making it capable of broadband transmission. DSL comes in many flavors, known collectively as xDSL. *See* ADSL, VDSL.
DTT	Digital Terrestrial Television (e.g., Ondigital, NTL).
DTV	Digital Television—The term adopted by the FCC to describe its specification for the next generation of broadcast television transmissions. DTV encompasses both HDTV and STV. *See* ATV, HDTV, STV.
D-Cab	Digital Cable Television (e.g., Telewest and NTL).
D-Sat	Digital Satellite Television (e.g., Sky).

DVB	Digital Video Broadcasting—A consortium of about 300 companies in the fields of broadcasting, manufacturing, network operation, and regulatory matters that have come together to establish common international standards for the move from analog to digital broadcasting. DVB is the organisation behind deployment and research into the creation of a single standard (API) for cross-platform interactive TV application creation. This European-centric "Holy Grail" is commonly known as DVB-MHP or Multimedia Home Platform.
DVD	Originally, digital video disk or digital versatile disk; now stands for nothing. Physically similar to a CD, a DVD is much more densely packed with data. Eventually it will contain the equivalent of eight hours of TV programming.
EPG	Electronic Programme Guide—An essential navigational device allowing the user to search for a particular program by theme or by category. They are currently the highest traffic areas on all platforms and are therefore of great interest to advertisers.
e-mail	Text messages created and viewed on PCs and transmitted electronically, usually over an office network or over the Internet.
Enhanced TV	Traditional TV programming or advertising which has an interactive element included within the broadcast stream. The distinction with interactive television is that interaction is not carried out within the broadcast stream and usually takes place within a walled garden.
Ethernet	A protocol for transmitting computer data over local area networks.
Fiber-Optic Cable	A cable containing one or more optical fiber strands. Each strand is capable in theory of carrying 25 trillion bits per second.
Fiber-Optics	Thin, flexible glass fiber cabling which is capable of handling large amounts of information/data. Fiber-optic cables provide the network to carry digital cable services.
FTP	File Transfer Protocol—A procedure for transmitting files of computer data over the Internet.

GEO	Geosynchronous communication satellite in the Clarke orbit at an altitude of 22,300 miles. It remains in a fixed position relative to the Earth.
Gigahertz	One billion hertz (q.v.). 28 and 38 gigahertz: portions of the spectrum designated by the FCC for terrestrial broadband fixed services. 28 gigahertz is currently used for LMDS; 38 gigahertz for wireless trunks.
HDTV	High Definition TV—A television that offers a very high-quality picture (similar to 35mm film) and sound (similar to audio CDs). HDTV uses digital rather than analog signal transmission. *See* DTV.
Headend	This is the electronic control center of a television system that processes the signal for transmission to digital and analog subscribers.
Hertz	The frequency in cycles per second of a wave form or carrier used for communication.
HFC	Hybrid Fiber-Coaxial System—A local cable TV or telephone distribution network consisting of fiber-optic trunks ending at neighborhood nodes, with coaxial cable feeder and drop lines downstream of the nodes.
HTML	Hyper Text Markup Language—A simple form of programming language used to build Web pages and simple interactive content on the NTL and Telewest digital cable platform.
HTTP	Hyper Text Transfer Protocol—Standard for transferring documents on the World Wide Web.
Hypermedia	A non-linear representation of information that allows users to access related works or images from a single computer screen. For example, a user reading an encyclopedia entry on jazz music can also hear excerpts from recordings and view photos of various artists. Sometimes synonymous with "multimedia."
Hypertext/Hyperlink	This is used to describe an action where a user can jump (by pressing a key on a keyboard or by pressing a button on a remote control) from one area to another area.
Information	In communication theory, a measure of one's freedom of choice in selecting a message, or of the range of possible

alternatives when receiving a message. The greater the number of potential messages, the greater the information contained in any one. Not to be confused with data or meaning.

Impressions   The number of times a banner, page, or page element on an iTV site is downloaded into the set top box, or the number of times a banner, page, or page element is seen by the viewer. The page of an interactive TV campaign can consist of several frames or graphics, and each of these elements will generate an impression. Accurate reporting of how many times a viewer saw a page can be difficult. In addition, caching issues affect impression rates.

Internet   The physical connections through which millions of computer users exchange data. The Internet comprises thousands of smaller networks, each associated with an organization such as firm, a university, a government agency, or an ISP. Communication is possible because of voluntary agreements to use certain communication techniques. *See* WWW.

Intranet   Interconnected IP networks confined within an organization, enterprise, or membership group; Intranets may be connected to the Internet.

Interoperability   Term describing the compatibility of content across all the platforms, i.e., it does not need to be repurposed in order to be viewed on all the platforms. *See* MHP.

IDTV   Abbreviation for Interactive Digital Television. This term is also used as "Integrated Digital Television," which represents a TV with a built-in STB.

Impulse Response   When a user presses a button on their remote or on their keyboard, they are immediately connected to the broadcaster. The modem (cable modem) which provides the access to the service is always "on" and requires no dial-up facility.

ISDN   Integrated Service Digital Network—An early and limited version of a digital subscriber line with capacity of either 64 or 128 kilobits per second. ISDN is viewed by some as a technological bridge between the current telephone system and an updated broadband network. Others see ISDN as a symbol of the failure of local telephone companies to adapt promptly to new technology.

ISP	Internet Service Provider—An organization that arranges connections between the Internet and individuals or enterprises. Large ISPs operate their own Internet backbones. Value-added ISPs (AOL, CompuServe) offer information services as well as interconnections.
ITC	Independent Television Commission—The ITC receives its powers through the 1990 and 1996 broadcasting acts. The ITC controls the issue of commercial broadcast licenses which govern the standard of programs and advertising. The ITC also acts as a watchdog body, regulating broadcasters and ensuring compliance with its regulations on program content, advertising, sponsorship, and technical performance.
iTV	Abbreviation for Interactive Television.
Ka-Band	A portion of the electromagnetic spectrum reserved by the FCC for both terrestrial and satellite uses, for which various specific proposals have been made.
Ku-Band	A portion of the electromagnetic spectrum designated by the FCC for, among other things, direct broadcast satellites.
LAN	Local Area Network—Communication paths linking computers, printers, and servers into a network for use by an individual, office, school, or other organization.
LMDS	Local Multichannel Distribution Service "wireless cable"—A new broadband wireless service operating in a frequency range (28 gigahertz) designated by the FCC for that purpose.
Local Loop	The pair of dedicated copper wires (or equivalent channel) that connects each telephone to a local switch or "central office." *See* Twisted Pair.
Metadata	The interactive component or data of a broadcast stream.
MHEG	Multimedia and Hypermedia Experts Group—A programming language used to build interactive TV applications for digital terrestrial TV in the UK. For example, all iTV applications for the ONdigital platform are built in MHEG.
MHP	Also known as DVB-MHP; developed as a middleware standard to develop content across all digital TV platforms. MHP standards have been determined by over 300 companies consisting of Philips and Microsoft, and will be implemented in Finland.

Modem	Modulator/demodulator—A device that transforms digital information into analog form for transmission over analog telephone lines, reversing the process for received data.
Moore's Law	The quantity of microelectronic processing speed, power, or memory that can be purchased with a dollar doubles every two years or so. In contrast, Internet traffic doubles two to five times per year. Named for Gordon Moore of Intel.
MMDS	Multichannel Multipoint Distribution System "wireless cable"—A local wireless terrestrial video broadcast technology that relies on line-of-sight transmission. An analog MMDS has up to 33 television channels; a digital MMDS may have 150 or more.
MPEG	Motion Picture Experts Group—An ISO-related industry standards organization that develops standards for coding video transmission.
MSO	Multiple System Operator—Any company that owns a large number of cable television systems.
Multicast	(1) A procedure for minimizing the Internet backbone capacity requirements of broadcasting identical simultaneous bitstreams to multiple recipients. (2) The use of digital spectrum assignments by broadcasters to air multiple channels or "standard" television rather than one channel of HDTV.
Multiplex	This is an ultra-high-frequency channel that is used to carry digital signals. Using compression, several services can be carried on the same frequency channel.
Multiplex Operator	A term for a broadcaster transmitting on one or more multiplexes. The UK digital terrestrial platform has four operators, BBC, Digital 3 & 4 (ITV and Channel 4), SDN. and Ondigital. Ondigital has three multiplexes, the others operate on one each. Multiplex operators do not necessarily have to be the originators of broadcast content, as Ondigital obtains its content from other providers.
Narrowband	A low-capacity communications link, such as a telephone cable, which with present technology is incapable of transmitting multiple TV signals. *See* Broadband.
Network	The collection of links that connects end users with one another and with devices such as servers, switches, and routers.

NTSC	National Television Standard Committee—Used to refer to the technical standards and physical characteristics of conventional analog TV broadcasting, as enshrined in FCC regulations.
NVOD	Near-Video-On-Demand—A facility deployed by digital satellite and cable operators. Movies are provided on multiple channels with staggered start times (i.e., every 10 minutes). Subscribers can then tune in to the next available showing. This is not available on digital terrestrial due to bandwidth constraints.
Platform	Term used to describe the distinct digital and analog distribution methods. For example, interactive TV applications are distributed across the various platform providers. The main platform providers include Telewest, NTL, Open, and BT (ADSL).
PC	Personal Computer.
PCS	Personal Communication Services—An FCC term for digital cell phones' recently auctioned frequencies.
POTS	Plain Old Telephone Service.
Protocol	A formal description of the message formats and rules that computers, switches, or other devices must follow when exchanging messages.
PPV	Pay-Per-View—Typically associated with a film channel or live sporting event where the viewer pays to watch. The same model as hiring a video from the video shop, but without the car journey and the inevitable late return.
Pay/Basic Ratio	A measure of the number of subscriptions to the pay channels as a percentage of total subscribers. As defined by the ITC in cable homes.
PVR	Personal Video Recorder—A separate unit that resembles a set top box and is plugged into the viewer's TV system. This intelligent device tracks and learns viewing preferences, seeks programs that meet viewer criteria, and automatically saves them for later retrieval. The requested content is digitally recorded on a hard disk drive located in the PVR in an MPEG2 format while also permitting the viewer VCR type control and functionality over the broadcast stream. The dominant platforms at present are TiVO and ReplayTV, although Sky Digital

will be integrating NDS's XTV PVR into their next generation of set top boxes.

Re-Versioned — The adaptation or modification of a site or service to allow its distribution or display on another media format or channel other than the one it was originally created for. Another commonly used term for this is "re-purposed."

Search Engine — Software that facilitates the discovery of relevant information in distant databases.

Shannon's Law — The quantity of accurate information that can flow over a channel has an upper limit determined by the bandwidth of the channel and its signal-to-noise ratio. Named for Claude Shannon of Bell Labs.

SMATV — Satellite Master Antenna Television—A private cable television system serving an apartment complex or similar residential grouping. Such systems currently serve around one million United States households.

Spectrum — The range of wavelengths (or frequencies) of electromagnetic radiation, from the longest radio waves to the shortest gamma rays. Visible light is only a small part of this range. "Frequency spectrum" or "airwaves."

STB — Set Top Box—Used to decode digital and analog transmissions for display on TVs. Also referred to as an "intelligent" box that connects to the TV set and also to the network which feeds the broadcast signal to home. Capability of these boxes is somewhat limited at present, due to the fact that they were given away to the consumer to encourage uptake. Use of more powerful, capable STBs will become more widespread as the platform providers increase their services.

Streaming — A real-time bitstream conveying audio or video information.

Switch-Off — Phrase used to describe the eventual hand-over from analog broadcasting to digital only.

SSL — Secure Socket Layer—A protocol developed by Netscape Communications Corporation which provides a means of encrypting information (like credit card details) over a network allowing for secure transactions. 128 bit is what most companies consider acceptable and is currently deployed at Telewest and Open Interactive.

Teletext	Commercial text service. The digital offering from Teletext is available on a new dedicated channel (Channel 9) on Ondigital, digital cable, and mobile services.
Transponder	One of the (12 or more) active electronic units in a communication satellite. Receives signals from Earth, translates to a different frequency, amplifies, and then broadcasts (downlinks) them.
Trunk	A physical or wireless broadband connection linking switches or routers to one another. Each trunk carries many connections. In contrast, a local loop generally connects a single subscriber with a nearby central office switch.
TVR	Television Rating—This is an expression of the number of times that a commercial has had the opportunity of being seen as a proportion of the number of people who make up the total audience. A rating or TVR is a percentage and is expressed in terms of a particular target audience.
Tiering	Used to describe the selling of channel packages to Pay TV subscribers.
Trigger	The point in a broadcast program when a viewer presses a button on the remote control and exits the broadcast stream to access interactive content.
UHF	Ultra High Frequency—A range of spectrum designated by the FCC for television broadcasts; originally channels 14 to 84, later reduced to 14 to 69.
Universe	The total number of people in a particular grouping, i.e., the UK TV universe refers to everyone who has a TV set in the UK.
URL	User Resource Location—Commonly referred to as an online address. For example: www.bmptvi.co.uk.
VCR	Videocassette recorder.
VHF	Very High Frequency—The original FCC-designated television broadcast spectrum. Channels 2 to 13.
Video-On-Demand/VOD	The user can watch a film or TV program when they want to watch it (on-demand). The broadcasters will offer consumers on-demand films initially, but may extend the service to all TV programs in the future. Obviously this is dependent upon

	many things. One of the main issues is that on-demand services are delivered via ADSL and the current penetration of ADSL is low. Building the infrastructure to deliver ADSL on a national scale is a huge undertaking from a technical and commercial perspective.
Web Page	An element of the interface offered to the user by a distant database, as displayed on the user's computer monitor when running a Web browser program.
Wide-Screen	Wide-screen television sets allows for wide-screen broadcasts to be displayed in a 16x9 format. All digital TV transmissions are broadcast in the wide-screen format. It is worth mentioning that the consumer does not need to buy a wide-screen TV to view digital TV (although you do need a set top box); a normal 4x3 aspect TV is fine.
Walled Garden	Term used to describe an area where the content is owned and controlled by the platform provider. All of the content is designed for the TV. A good example of a walled garden on the Internet is AOL.
Wireless Cable	*See* Gigahertz; LMDS; MMDS. Wireless cable systems of all kinds served fewer than one million United States households in 1997.
xDSL	*See* DSL; ADSL.
Zapping	Switching from channel to channel, normally via a remote control and often during a commercial break.
Zipping	Fast forwarding through a commercial break while playing on a VCR.

References

Aaker, D.A. (1991). *Managing Brand Equity, Capitalizing on the Value of a Brand Name.* New York: The Free Press.

Abe, G. (1997). *Residential Broadband.*, Indianapolis, IN: Cisco Press.

Abell, D. (1980). *Defining the Business: The Starting Point of Strategic Planning.* Englewood Cliffs, NJ: Prentice-Hall.

Abell, D. (1994). *Managing Dual Strategy.* New York: The Free Press.

Access Conference International. (2001). *The Interactive TV Show USA.* New York, Essex House.

Adams, W. & Yellen, J.L. (1976). Commodity bundling and the burden of monopoly. *Quarterly Journal of Economics,* 90, 475-498.

AIM—Associazione Interessi Metropolitani & RESEAU. (1996). *Progetto Milano per la Multimedialità— Gli Scenari Tecnologici.* Milan, Italy, Quaderni AIM, Volume 1.

Ajello, B. (1998). Le principali tendenze della TV digitale in Europa e in Italia. In Istituto Economia dei Media & Fondazione Rosselli (Eds.), *L'Industria Della Comunicazione in Italia.* Turin, Italy: Guerini e Associati.

Alattar, A. M. & Adnan, M. (2000). Smart images using Digimarc's watermarking technology. *Proceedings of the IS&T/SPIE's 12th International Symposium on Electronic Imaging,* San Jose, California, January 25. Volume 3971, Number 25.

Albers, S., Bachem, C., Clement, M. & Peters, K. (1998). Marketing instrumente. Produkte und Inhaltinhalt. In Albers, S., Clement, M. & Peters, K. (Eds.), *Marketing mit Interaktiven Medien.* Frankfurt, Germany: a.M., IMK, pp. 267-282.

Ancarani, F. (1999). *Concorrenza e Analisi Competitiva*. Milan, Italy: EGEA.

Anderson, P. & Tushman, M.L. (1990). Technological discontinuities and dominant designs: A cyclical model of technological change. *Administrative Science Quarterly*, 35, 604-633.

Ang, I. (1991). *Desperately Seeking the Audience*. London: Routledge.

Antonelli, C. (1982). *Cambiamento Tecnologico e Teoria di Impresa*. Turin, Italy: Loescher.

ARD (1999). *Media Perspektiven Basisdaten*. Frankfurt, Germany: a.M. IMK.

Association of American Publishers. (2000). *Digital Rights Management for Ebooks: Publisher Requirements*. Washington, DC: Association of American Publishers.

Association of American Publishers Copyright Committee. (2000). *Contractual Licensing, Technological Measures and Copyright Law*. Washington, DC: Association of American Publishers.

Association of American Publishers Rights and Permissions Advisory Committee. (2000). *The New & Updated Copyright Primer: A Survival Guide to Copyright and the Permissions Process*. Washington, DC: Association of American Publishers.

Bailer, B. (1997). *Geschäftsmodelle: Methoden und Qualität*. PhD Dissertation, Department of Computer Science, University of Zurich, Switzerland.

Bane, W., Bradley, S. & Collis, D. (1997). Winners and losers: Industry structure in the converging world of telecommunications, computing and entertainment. In Yoffie, D.B. (Ed.), *Competing in the Age of Digital Convergence*. Boston, MA: Harvard Business School Press.

Baran, P. (1962). *On Distributed Communications Networks*. Santa Monica, CA: RAND Corporation.

Barbero, M., Cucchi, S. & Stroppiana, M. (1991). A bit reduction system for HDTV transmission. *IEEE Transactions on Circuits and Systems for Video Technology*, 1(1), 4.

Barbero, M. & Stroppiana, M. (1994). Video compression techniques and multilevel approach. *SMPTE Journal*, 103(5).

Barney, J.B. (1991). Firm resources and sustained competitive advantage. *Journal of Management*, 17(1), 99-120.

Barnow, E.A. (1970). *History of Broadcasting in the United States*. New York: Oxford University Press.

Bartholomew, M.F. (1997). *Successful Business Strategies Using Telecommunications Services.* Boston, London: Artech House.

Beebe, J. (1977). Institutional structure and program choice in television markets. *Quarterly Journal of Economics,* 91.

Besen, S.M. & Saloner, G. (1989). The economics of telecommunications standard. In Crandall, R. & Flamm, K. (Eds.), *Changing the Rules: Technological Change, International Competition, and Regulation in Communications.* Washington, DC, Brookings Institution.

Besen, S.M. (1987). *New Technologies and Intellectual Property: An Economic Analysis.* Note N-2601-NSF, Santa Monica, CA: RAND Corporation, May.

Bienert, P. (2000). Das mediengerät der zukunft—Fernsehcomputer oder computerfernseher? *HMD,* 211, 43-52.

Blaxter, L., Hughes, C. & Tight, M. (1996). *How to Research.* Philadelphia: Open University Press.

Blödorn, S., Gerhards, M. & Klingler, W. (2000). Fernsehen im neuen jahrtausend—Ein informationsmedium? *Media Perspektiven,* 4, 171-180.

Boni, M. (1997). *Modelli di Business Nell'era Digitale. Information Market e Risposte Operative.* Milan, Italy: Guerini e Associati.

Bordewijk, J. & Kaam, B. (1986). Towards a new classification of teleinformation services. *Inter Media,* 14, 1.

Bower, J.L. (1970). *Managing the Resource Allocation Process : A Study of Corporate Planning and Investment.* Boston, MA: Harvard Business School Press.

Bowler, J. (2000). DTV content exploitation. What does it entail and where do I start? *New TV Strategies,* 2(7), 7.

Bradley, S., Hausman, J. & Nolan, R. (1993). *Globalization, Technology and Competition. The Fusion of Computers and Telecommunications in the 1990s.* Boston, MA: Harvard Business School Press.

Brinkley, J. (1997). *Defining Vision: The Battle for the Future of Television.* New York: Harcourt Brace and Co.

Broadcasting Act. (1996). *The Parliamentary Office.* London: HMSO.

Brown, S.L. & Eisenhardt, F.M. (1999). *Competing on the Edge.* Boston, MA: Harvard Business School Press.

Brusio, G. (1993). *Economia e Finanza Pubblica*. Rome, Italy: NIS La Nuova Italia.

Burns, C. (1995). *Copyright Management and the NII: Report to the Enabling Technologies*. Washington, DC: Association of American Publishers.

Busacca, B. & Troilo, G. (1992). La diffusività intersettoriale dell'immagine di marca. *Economia & Management,* 5, 75-76.

Busacca, B. & Valdani, E. (1992). Customer satisfaction: Una nuova sfida. *Economia & Management,* 2.

Busacca, B. (1990). *L'Analisi del Consumatore. Sviluppi Concettuali e Implicazioni di Marketing.* Milan, Italy: EGEA.

Busacca, B. (1994). Le risorse aziendali customer based. Potenziale generativo e condizioni di sviluppo. *Economia & Management,* 5.

Busacca, B. (1994). *Le Risorse di Fiducia dell'Impresa.* Turin, Italy: UTET.

Busacca, B. (1995). Le strategie di brand extension: L'attivazione del valore-potenzialità della marca. In Adams, P., Bertoli, G., Busacca, B., Gnecchi, M., Mazzei, R., Verona, G. & Vicari, S. (Eds.), *Brand Equity: Il Potenziale Generativo della Fiducia.* Milan, Italy: EGEA, pp. 157-198.

Campobasso, G.F. (1991). *Diritto Commerciale.* Turin, Italy: UTET.

Carey, J. (1989). *Interactive Media, International Encyclopedia of Communications.* New York: Oxford University Press.

Castaldo, S. & Verona, G. (1998). *Lo Sviluppo di Nuovi Prodotti.* Milan, Italy: EGEA.

Champy, J. (1995). *Reengineering Management.* New York: Harper Collins Publishers Inc.

Chandler, A. (1990). *Scale and Scope.* Cambridge, MA: Harvard University Press.

Clark, J. (1995). *Managing Innovation and Change.* Newbury Park: Sage Publications.

Coase, R.H. (1950). *British Broadcasting: A Study in Monopoly.* Cambridge, MA: Harvard University Press.

Collins, D.J., Bane, W. & Bradley, S.P. (1997). Industry structure in the converging world of telecommunications computing and entertainment. In Yoffie, D.B. (Ed.), *Competing in the Age of Digital Convergence.* Boston, MA: Harvard Business School Press, pp. 159-201.

Convergent Decision Group. (1998). *Digital Terrestrial Television in Europe.* Convergent Decision Group Report, London, UK.

Costabile, M. (1996). *Misurare il Valore per il Cliente.* Turin, Italy: UTET.

Cozzi, G. & Molinari, M. (1990). L'immagine di marca: Come costruirla, come gestirla, come modificarla. *Economia e Diritto del Terziario,* 2.

Cronin, M.J. (1994). *Doing Business on the Internet: How the Electronic Highway Is Transforming American Companies,* New York: Van Nostrand Reinhold.

Dameri, R.P. (1996). Analisi strategica mediante scenari. *Sinergie,* 39, 193-210.

Datamonitor. (1999). *Interactive TV Markets in Europe and the U.S. 1998-2003.* Report (Executive Summary).

Datamonitor. (2001). *Is the Channel Dead? The Impact of Interactivity on the TV Industry.* Available online at: www.datamonitor.com.

Davenport, H. (1998). Better than expected "coverage of ONDigital." *New Media Market,* 16(43).

Davenport, T.H. (1993). *Process Innovation.* Boston, MA: Harvard Business School Press.

David, P.A. (1975). *Technical Choice, Innovation and Economic Growth: Essays on America and British Experience in the Nineteenth Century.* Cambridge, MA: Cambridge University Press.

Dematté, C. & Perretti, F. (1997). *L'Impresa Televisiva.* Milan, Italy: ETAS.

Dematté, C. (1995). Le autostrade elettroniche cambieranno il modo di vivere e l'intera economia. *Economia & Management,* 2.

Di Bernardo, B. & Rullani, E. (1984). Evoluzione: Un nuovo paradigma per la teoria dell'impresa e del cambiamento tecnologico. *Economia e Politica Industriale, volume.*

Di Bernardo, B. & Rullani, E. (1990). *Il Management e le Macchine.* Bologna, Italy: Il Mulino.

Dierick, I. & Cool K. (1989). Asset stock accumulation and sustainability of competitive advantage. *Management Science,* 35(12), 1504-1511.

Dittberner Associates, Inc. (1995). *Advanced Technologies Series Update,* Project ESS, N. 33.

Dowling, M., Lechner, C. & Thielmann, B. (1998). Convergence—Innovation and change of market structures between television and online services. *Electronic Markets Journal*, 8(4), 31-35.

Doyle, P. (1989). Building successful brands: The strategic options. *Journal of Marketing Management*, 5(1), 77-95.

Drucker, P.F. (1992). *Managing for the Future*. Dutton: Truman Talley Books.

DTI and DCMS. (1998). *Regulating Communications—Approaching Convergence in the Information Age*. London: HMSO.

Dunnett, P. (1990). *The World Television Industry*. New York: Routledge.

Durlak, J. (1987). A typology for interactive media. In McLaughlin, M. (Ed.), *Communication Yearbook 10*. Newbury Park: Sage Publications.

Eisenhardt, K.M. (1989). Building theories from case study research. *Academy of Management Review*, 14(4), 532-550.

EITO—European Technology Observatory (2001). *Annual Report*. Frankfurt, Germany: EITO.

EMNID TNS (1996). *Telematik in Haushalten*. Bielefeld: Große Haushaltsumfrage Deutschland.

European Commission. (1997). *Green Paper on the Convergence of the Telecommunications, Media and Information Sectors, and the Implications for Regulation. Towards an Information Society Approach*. Brussels.

European Communication Council. (1999). *Report Die Internet-Okonomie*, Berlin, Heidelberg, Germany, pp. 237-244.

European Institute for the Media. (1991). *The Future of High Definition Television in Europe*.

Eutelis Consult, Gaida, K. (1999). *Der Markt für Internet & TV-Dienste in USA und Deutschland*. Kurzstudie: Ratingen.

Farinet, A. (1995). *Tecnologia e Concorrenza nei Mercati Industriali*. Milan, Italy: EGEA.

FCC. (2001). *In the Matter of Non-Discrimination in the Distribution of Interactive Television Service over Cable*. CS Docket, No. 01-7, p.2.

Feldman, T. (1997). *An Introduction to Digital Media*. London: Routledge.

Ferguson, C. & Morris, C. (1993). *Computer Wars: How the West Can Win in a Post-IBM World.* New York: Random House.

Fiocca, R. & Corvi, E. (1996). *Comunicazione e Valore nelle Relazioni d'Impresa.* Milan, Italy: EGEA.

Flynn, B. (2000). *Digital TV, Internet & Mobile Convergence—Developments and Projections for Europe.* Digiscope Report, London: Phillips Global Media.

Forrester. (1998). *Europe's Internet Growth.* Forrester Report.

Forrester. (1998). *Lazy Interactive TV.* Forrester Research.

Forrester. (2000). *Europe I-DTV Walls Come Down.* Forrester Research.

Fournier, G.M. (1986). The determinants of economic rents in television broadcasting. *Antitrust Bulletin,* 31, 1045-1066.

Fowler, M.S. & Brenner, D.L. (1982). A marketplace approach to broadcast regulation. *Texas Law Review,* 60, 207-257.

Gambaro, M. & Silva, F. (1992). *Economia della Televisione.* Bologna, Italy: Il Mulino.

Gandy, O.H. Jr. (2000). Race, ethnicity and the segmentation of media markets. In Curran, J. & Gurevitch, M. (Eds.), *Mass Media and Society.* New York: Oxford University Press., pp. 44-69.

Garcia-Murillo, M. & McInness, I. (1998). Will Internet success lead to broadband failure? *Proceedings of the CRTPS 4th Annual Conference,* Ann Arbor, Michigan, 5/6, 6.

GFK-Medienforschung: Spohrer, M. (1998). *Mediennutzung im Wandel—TV und PC?* Lecture to the Euroforum Conference, Hamburg, Germany.

Goertz, L. (1995). Wie interaktiv sind medien? *Rundfunk und Fernsehen,* 4.

Goodall and Associates. (1996). *Scenarios for Digital Television.* Mimeo.

Grant, A.E. & Shamp, S.A. (1997). Will TV and PCs converge? Point and counterpoint. *New Telecom Quarterly,* (2nd Q.), 31-37.

Grauer, M. & Merten U. (1996). *Multimedia.* Berlin, Heidelberg et al.: Springer.

Greenstein, S. & Khanna, T. (1997). What does convergence mean? In Yoffie D.B. (Ed.), *Competing in the Age of Digital Convergence.* Boston, MA: Harvard Business School Free Press, pp. 201-225.

Greenstein, S. M. (1999). Industrial Convergence. In Dorf, R. (Ed.), *The Technology Management Handbook*, Boca Raton, FL: CRC Press.

Habann, F. (2000). Management of core resources: The case of media enterprises. *Journal of Media Management*, 2(1).

Hamel, G. & Prahalad, C.K. (1994). *Competing for the Future*. Boston, MA: Harvard Business School Press.

Hamel, G. & Prahalad, C.K. (1994). *Competing for the Future: Breakthrough Strategies for Seizing Control of Your Industry and Creating the Markets of Tomorrow*. Boston, MA: Harvard Business School Free Press.

Hamermesh, R. (1986). *Making Strategy Work*. New York: John Wiley & Sons.

Hannan, M.T. & Freeman, J. (1984). Structural inertia and organizational change. *American Sociological Review*, 49, 149-164.

Hansen, G. & Wernerfelt, B. (1989). Determinants of firm performance. The relative importance of economic and organizational factors. *Strategic Management Journal*, 10, 399-511.

Heeter, C. (1989). Implications of new interactive technologies for conceptualizing communication. In Salvaggio, J. & Bryant, J. (Eds.), *Media Use in the Information Age: Emerging Patterns of Adoption and Consumer Use*. Hillsdale, NJ: Lawrence Erlbaum Associates.

Heil, B. (1999). *Online-Dienste, Portal Sites und Elektronische Einkaufszentren*. Wiesbaden: Gabler Edition Wissenschaft.

Herring, J.M. & Gross G.C. (1936). *Telecommunications: Economics and Regulation*. New York: McGraw Hill.

Hess, T. & Herwing, V. (1999). Portale im Internet. *Wirtschaftsinformatik*, 6, 551-553.

Homeyer, J. & Peters, R.H. (1999). Microsoft—Zäher start. *Wirtschaftswoche*, 4, 44-45.

Iannella, R. (2001). Digital Rights Management (DRM) architectures. *D-Lib Magazine*, 7(6).

Intermedia. (1999). *White Paper on Interactive TV*. Intermedia Report.

Intermedia Special Report. (1999). *Digital Terrestrial TV: Descrambling the issues*. 25(9).

Istituto Economia dei Media & Fondazione Rosselli. (1998). *L'Industria della Comunicazione in Italia*. Milan, Italy: Guerini e Associati.

ITC. (1997). *ITC Announces its Decision to Award Multiple Service Licences for Digital Terrestrial Television.* London: ITC Press Release.

Johnson, L.L., & Castleman, D.R. (1991). *Direct Broadcast Satellites: A Competitive Alternative to Cable Television?* Santa Monica, CA: RAND Corporation.

Kahin, B. & Kate, A. (Eds.). (1996). Forum on technology-based intellectual property management: Electronic commerce for content. *Interactive Multimedia News,* 2(Special Issue).

Katz, H. (2000). Interactivity 2000: An industry viewpoint. *Journal of Interactive Advertising,* 1(1), Fall.

Koch, R. (2000). *The Financial Times Guide to Strategy: How to Create and Deliver a Useful Strategy.* London: Prentice Hall-Pearson Education.

Laurel, B. (1991). *Computers as Theatre.* Reading, MA: Addison-Wesley.

Lessig, L. (2001). *The Future of Ideas: The Fate of the Commons in a Connected World.* New York: Random House.

Lichtenberg, L. (1999). Influences of electronic developments on the role of editors and publishers, strategic issues. *Journal of Media Management,* (1), 23-30.

Liscia, R. (Ed.). (1998). *Il Mercato dell'Editoria Multimediale.* Milan, Italy: Guerini e Associati.

Litman, J. (2001). *Digital Copyright.* Amherst, NY: Prometheus Books.

Lyon, G. (2001). *The Internet Marketplace and Digital Rights Management.* Broadway: National Institute for Standards and Technology.

MacDonald, J. & Tobin, J. (1998). Customer empowerment in the digital economy. In Tapscott, D. (Ed.), *Blueprint Digital Economy.* New York: McGraw-Hill, pp. 202-220.

Mandelli, A. (1998). La nuova televisione interattiva: L'incontro tra Internet e la TV. *Economia & Management,* (2).

Marra, A. (1998). *La Televisione Digitale nel Contesto Internazionale.* Bologna, Italy: Il Mulino, pp. 161-176.

Mauri, C. (1990). *Concorrenza Dinamica.* Milan, Italy: EGEA.

Mayer, M., Mohn, W. & Zabbal, C. (2001). *PCs vs. TVs.* Accessed online, August 30, 2001, at: http://mckinseyquarterly.com/.

McLuhan, M. (1964). *Understanding Media: The Extensions of Man.* New York: McGraw Hill.

McLaughlin, J. (1985). Mapping the information business. *Program on Information Resource Policy.* Boston, MA: Harvard University Press.

Ministero delle Poste e Telecomunicazioni. (1996). *Report of the Work Group: Study on the Application Possibilities of MMDS/LMDS Technologies to Microwave Distribution of Television Services, and More Generally, New Multimedia Services.* Rome, Italy: Ministry for PT Italy.

Moore, J. F. (1996). *The Death of Competition.* New York: John Wiley & Sons.

Moore, G. A. (1990). *Marketing and Selling High-Tech Products to Mainstream Customers*, London, UK: Harper Collins Publishers.

Murroni, C. & Irvine, N. (1998). *Access Matters.* London: IPPR.

Murroni, C. (1999). Gli sviluppi nel regno unito. In Quaderni Istituto Economia dei Media (Ed.), *La Televisione Digitale Terrestre: Tendenze di Sviluppo, Vantaggi e Problemi.* No. 9, Milan, Italy: Fondazione Rosselli, pp. 32-43.

Napoli, P. (2001). The audience product and the new media environment: Implications for the economics of media industries. *Media Journal,* 3(II).

Negroponte, N. (1995). *Being Digital.* New York: Alfred A. Knopf.

Neuman, W.R. (1992). The technological convergence: Television networks and telephone networks. *Television in the 21st Century: The Next Wave.* Aspen Institute Program on Communications and Society.

Noelle-Neumann, E., Schulz, W. & Wilke, J. (1999). *Publizistik Massenkommunikation.* Frankfurt, Germany: a.M., Fischer.

Normann, R. & Ramìrez, R. (1995). *Le Strategie Interattive d'impresa: Dalla Catena alla Costellazione del Valore.* Milan, Italy: ETAS Libri.

OECD. (ed.). (1992). *Telecommunications and Broadcasting—Convergence or Collision?* Paris, France: OECD.

Oftel. (2000). *International Benchmarking of DSL Services.* London: Oftel Report.

Owen, B.M. (1974). Organisation of the television industry. In Owen B.M., Beebe, J.H. & Manning, W.G. *Television Economics.* Lexington, MA: Heath.

Owen, B.M. (1999). *The Internet Challenge to Television.* Cambridge, MA: Harvard Press.

Pagani, M. (1999). Interactive television: A model of analysis of business economic dynamics. *Proceedings of the 6th Symposium on Emerging Electronic Market,* Muenster, Germany.

Pagani, M. (2001). Content management for a digital broadcaster. *Proceedings of Managing Information Technology in a Global Economy, 2001 IRMA Conference,* pp. 1062-1066.

Pagani, M. (2002). Measuring the potential for IT convergence at macro level: A definition based on platform penetration and CRM potential. *Proceedings of Issues & Trends of Information Technology Management in Contemporary Organizations, 2002 IRMA Conference,* pp. 1060-1063.

Pagani, M., (2000). Interactive television: A model of analysis of business economic dynamics. *Journal of Media Management,* 2(I), 25-37.

Pagani, M., (2000). *La TV nell'era Digitale: Le Nuove Frontiere Tecnologiche e di Marketing della Comunicazione Televisiva.* Milan, Italy: EGEA.

Picard, R.G. (2000). *Changing Business Models of Online Content Services: Their Implications for Multimedia and Other Content Producers. Journal of Media Management,* 2(II).

Pilati, A. (1999). Alternative d'uso nel disegno dei nuovi sistemi ad alta capacità. *Quaderni dell'Istituto di Economia dei Media,* 9(July).

Pilati, A. (Ed.). (1993). *L'Industria della Comunicazione in Europa.* Rome, Italy: Sipi.

Pine, B.J. (1992). *Mass Customization: The New Frontier in Business Competition.* New York: Harvard Business School Free Press.

Piol, E. (1995). *Fibra Ottica e Multimedialità Interattiva.* Parigi, France, European IT Forum.

Podesta, S. (1988). Teoria dell'impresa ed economia industriale. In Guatri, L. (Ed.), *Trattato di Economia delle Aziende Industriali.* Milano, Italy: EGEA.

Poltrak, D. (1983). *Television Marketing.* New York, NY: McGraw Hill.

Porat, M. (1981). *The Information Economy,* Institute for Communication Research, Stanford University, Stanford.

Porter, M.E. (1990). *The Competitive Advantage of Nations.* New York: The Free Press.

Porter, M.E. (1980). *Competitive Strategy.* New York: The Free Press.

Porter, M.E. (1985). *Competitive Advantage: Creating and Sustaining Superior Performance.* New York: The Free Press.

Postman, N. (1992). *Technolopoly.* New York: First Vintage Books.

Prahalad, C.K. & Hamel, G. (1990). The core competence of the corporation. *Harvard Business Review,* 68(3), 79-91.

Pulcini, E. (1999). *Dopo Internet: Storia del Futuro dei Media Interattivi.* Rome, Italy: Castelvecchi.

Rawolle, J. & Hess, T. (2000). New digital media and devices—An analysis for the media industry. *Journal of Media Management,* 2(2), 89-98.

Richeri, G. (1993). Gli aspetti economici dell'alta definizione. In Istituto Economia dei Media (Ed.), *Mind-L'Industria della Comunicazione in Europa.* Rome, Italy: Sipi.

Richeri, G. (1993). *La TV che Conta. Televisione Come Impresa.* Bologna, Italy: Baskerville.

Richeri, G. (1999). Reti di trasmissione digitali terrestri: Una strategia di lungo periodo. In Quaderni di Economia dei Media (Ed.), *La Televisione Digitale Terrestre: Tendenze di Sviluppo Vantaggi e Problemi.* Milan, Italy: Fondazione Rosselli.

Robinson, Godbey, & Geoffrey (1997). *Time For Life: The Surprising Ways Americans Use Their Time,* University Park, PA: Pennsylvania State University Press.

Rogers, E. (1986). *Communication Technology. The New Media in Society.* New York: The Free Press.

Rosenblatt, B. et al. (2001). *Digital Rights Management: Business and Technology.* New York: John Wiley & Sons.

Rosenblatt, B. (1996). *Two Sides of the Coin: Publishers' Requirements for Digital Intellectual Property Management.* Inter-Industry Forum on Technology-Based Intellectual Property Management, Washington, DC.

Rullani, E. & Vaccà, S. (1987). Scienza e tecnologia nello sviluppo industriale. *Economia e Politica Industriale,* (53).

Rullani, E. (1984). Teoria ed evoluzione dell'impresa industriale. In Rispoli M. (Ed.), *L'Impresa Industriale.* Bologna, Italy: Il Mulino.

Rullani, E. (1987). Alcune considerazioni sul metodo; un commento. *Finanza Marketing e Produzione,* (3).

Samuelson, P. (1999). *Intellectual Property in the Age of Universal Access.* New York: Association for Computing Machinery.

Saunders, M., Lewins, P. & Thornhill, A. (1997). *Research Methods for Business Students.* London: Pearson Professional Ltd.

Scalfi, G. (1986). *Manuale di Diritto Privato.* Turin, Italy: UTET.

Schumann, D., Artis, A. & Rivera, R. (2001). The future of advertising viewed through an IMC lens. *Journal of Interactive Advertising,* 1(2), Spring.

Schumpeter, J.A. (1934). *The Theory of Economic Development.* Cambridge, MA: Harper Bros.

Schumpeter, J.A. (1950). *Capitalism, Socialism and Democracy* (3rd Edition). New York: Scientifica.

Schumpeter, J.A. (1961). The theory of economic development. *Inquiry into Profit, Capital, Credit, Interest and the Business Cycle.* Cambridge, MA: Harvard University Press.

Shannon, C. E. & Weaver, W. (1949). *The Mathematical Model of Communication.* Urbana, IL: University of Illinois Press.

Shannon, C.E. (1948). The mathematical theory of communication. *Bell System Technical Journal,* (27), 379-423, 623-656.

Sherman, B.L. (1998). *Telecommunications Management.* New York: McGraw Hill.

Squire Sanders & Dempsey (1998). *Study on Adapting the EU Telecommunications Regulatory Framework to the Developing Multimedia Environment.*

Stefik, M. (1996). Letting loose the light: Igniting commerce in electronic publication. *Internet Dreams: Archetypes, Myths, and Metaphors.* Cambridge, MA: MIT Press.

Stewart, C. (1999). Minefield or goldmine: What can we expect from interactive TV? *New TV Strategies,* 1(1), 10-14.

Swain, P. & Blustin, A. (2000). *Personal Television Services: The Impact on Advertisers.* White Paper.

Szuprowicz, B. (1995). *Multimedia Networking:* New York: McGraw-Hill.

Tapscott, D. (1999). Privacy in the digital economy. In Tapscott, D. (Ed.), *The Digital Economy.* New York: McGraw Hill, p. 275s.

Teece, D.J. (1986). Innovazione tecnologica e successo imprenditoriale. *L'Industria,* (4), 605-643.

Telecommunications. (1997). *A Technology Guide*. London: BPA Group Ltd.

Todreas, T.M. (1999). *Value Creation and Branding in Television's Digital Age.* Westport, CN: Quorum Books.

Vaccà, S. (1998). Imprese e sistema industriale in una fase di rapida trasformazione tecnologica. In Guatri, L. (Ed.), *Trattato di Economia delle Aziende Industriali*. Milan, Italy: EGEA, pp. 78-79.

Valdani, E., Ancarani, F. & Castaldo, S. (2001). *Convergenza. Nuove Traiettorie per la Competizione*. Milan, Italy: EGEA.

Valdani, E. & Busacca, B. (1999). Customer base view. *Finanza Marketing e Produzione*, 2, 95-131.

Valdani, E. (1996). *Dall'Evoluzione alla Coevoluzione dalla Competizione all'Ipercompetizione.* Working Paper SDA Bocconi, Milan, Italy.

Valdani, E. (1996). *Marketing Strategico—Un'impresa Proattiva per Sviluppare Capacità Market Driving e Valore.* Milan, Italy: ETAS Libri.

Valdani, E. (2000). *L'impresa Pro-Attiva Co-Evolvere e COMpetere nell'era dell'Immaginazione.* Milan, Italy: McGraw-Hill.

Van Tassel, J. (2001). *Digital Content Management: Creating and Distributing Media Assets by Broadcasters.* Washington, DC: NAB Research and Planning Department, National Association of Broadcasters.

Vannucchi, G. (1998). *La Storia della Televisione Digitale.* Available online at: www.mediamente.rai.it.

Varian, H.R. (1995). Pricing information goods. Paper presented at *Research Libraries Group Symposium on Scholarship in the New Information Environment,* Harvard Law School, Cambridge, Massachusetts.

Vicari, S. (1991). *L'impresa Vivente.* Milan, Italy: ETAS Libri.

Vicari, S. (1998). *Le Nuove Dimensioni della Concorrenza.* Milan, Italy: EGEA.

Vogel, H.L. (1990). *Entertainment Industry Economics: A Guide for Financial Analysis* (2nd Edition). Cambridge, MA: Cambridge University Press.

Von Hippel, E. (1988). *The Sources of Innovation.* London: Basil Blackwell.

Webb, J.K. (1983). *The Economics of Cable Television.* Lexington, MA: Lexington Books.

Weiser, M. & Brown, J.S. (1998). Center and periphery. In Tapscott, D. (Ed.), *Blueprint Digital Economy.* New York: McGraw-Hill, pp. 316-35.

Werbach, K. (1997). *Digital Tornado: The Internet and Telecommunications Policy.* Washington, DC: FCC Office of Plans and Policy, Working Paper 29.

Wildman, S.S. & Owen, B.M. (1985). Program competition, diversity, and multichannel bundling in the new video industry. In Noam, E.M. (Ed.), *Video Media Competition: Regulation, Economics, and Technology.* New York: Columbia University Press.

Williamson, O. (1975). *Markets and Hierarchies, Analysis and Antitrust Implications: A Study in the Economics of Internal Organisation.* New York: The Free Press.

Wilson, K.G. (1988). *Technologies of Control: The New Interactive Media for the Home.* Madison, WI: University of Wisconsin Press.

Wirtz, B. W. (1999). Convergence processes, value costellations and integration strategies in the multimedia business, *The International Journal on Media Management, 1(1)*.

Wössner, M. (1999). *Vorlesung zur Medienwirtschaft.* St. Gallen, Switzerland: MCM Institute.

Yoffie, D.B. (1997). Introduction chess and competing in the age of digital convergence. In Yoffie, D.B. (Ed.), *Competing in the Age of Digital Convergence.* Boston, MA: Harvard Business School Free Press, pp. 1-36.

Yoffie, D.B. (Ed). (1997). *Competing in the Age of Digital Convergence.* Boston, MA: Harvard Business School Free Press.

Ziola, T. (1998). *Web TV Lifestyle—The Convergence of Digital TV and PC Technology.* Lecture to the Deutsche Telekom Multimedia Symposium, Virtual Life—Leben im Netz, Munchen, Germany.

About the Author

Margherita Pagani

Margherita Pagani is a Researcher at the I-LAB Research Center on Digital Economy at the Bocconi University in Milan, Italy, where she coordinates New Media & Tv-Lab. She is an Adjunct Professor in Management at the Bocconi University, Italy.

Dr. Pagani has written two books on digital interactive television and several papers on interactive television, digital convergence, and content management which have been discussed in many academic conferences in Europe and USA. She has worked with RAI Radiotelevisione Italiana and as a Member of the Workgroup "Digital Terrestrial" (Permanent Forum of Communications) for the Ministry of Communications in Italy.

Index

Symbols

3G Mobile Information Device (MID 3G) 36

A

access control (AC) 61
ADSL (Asymmetric Digital Subscriber Line) 59, 63, 79
ASL (Asymmetric Subscriber Loop) 57
alliances 17
application programming interface (API) 127
archive management 159
awareness aggregate 146

B

bandwidth 5, 56
barriers to convergence 37
billing services 190
brand awareness 136
brand communication tools 150
brand equity 135
brand identity 136
brand image 136
branding strategies 135
broadband multimedia PC 37
byte 5

C

cable 78
cable operators 16
capacity 110
CATV (cable networks) 61
chain competition 140
channel content management 158
channels for specific viewer-groups 139
clearing house services 190
communication model 33
conditional access (CA) 69, 126
consultation 106
consultational interactivity 108
consumer attitudes 39
content gatekeeper 129
content manipulation 8
content media management 156
content packaging 8
content processes 192
content production 8
content providers 15
content reception 8
content transmission 8
contract and rights management 189
convergence 1, 31
convergence factor 43
convergence index 47
conversation 105
conversational interactivity 108
cooperative paradigm 25
copyright 180

critical digital mass index 42
crossover competition 140
customer gatekeeper 129
customer relationship management 42

D

DAL—dedicated advertiser location 124
DBS (direct broadcast satellite) 57
decoder 67, 111
delivery management 159
diffusive systems 97
digital asset management 158
digital cable transmission 61
digital coding 54
digital compression 55
digital portals 163
digital rights management 180, 186
digital satellite transmission systems 60
digital signal 5, 54, 75, 160
digital technology 53, 75
digital television 53, 76
digital television networks types 138
digital television transmission 90
digital terrestrial television (DTT) 62, 75
digital video compression 55
digitalisation 5
direct to home (DTH) satellite systems 77
Disney channel 142
distribution costs 91
downstream 97
DSL 63
DTH (direct to home) 57
DVB—digital video broadcasting 71

E

economic effects of digital transmission 85
electromagnetic spectrum 57
Electronic Programming Guide (EPG) 70, 117, 147
end device 35, 126
enhanced TV 100
enterprise-wide content management 158

evoked aggregate 146

F

finance processes 192

G

general channels 139
general radio packet service (GRPS) 35
global system mobile communication (GSM) 34

H

high definition TV (HDTV) 58

I

ICT (information communication technology) 4
industry and firm dimension 32
innovations in Internet-TV convergence 28
intellectual property 181
intellectual property (IP) asset creation 187
intellectual property asset delivery management 188
intellectual property media asset management 188
intellectual property rights (IPRs) 181
interactive advertising 120
interactive digital television (iDTV) value chain 125
interactive digital television marketplace 125
interactive digital TV 197
interactive games 119
interactive multimedia services 98
interactive portal tools 152
interactive programming 123
interactive systems 97
interactive television (iTV) 96, 196
interactive transmission systems 97
interactivity 97
interactivity factor 46
International Standards Organization (ISO) 56

Internet 6
Internet-via-TV 113
intersectorial competition 141
IP multicast 65
IRD (integrated receiver decoder) 61

K

key drivers of convergence 4

L

LNB (low noise block) 61
legislation drivers 6
license management 190
local interactivity 100
local multichannel distribution service (LMDS) 83
local multipoint distribution system (LMDS) 67

M

media asset management (MAM) 159
Mediaset 170
metamarket 1
Metcalf's law 6
middleware 127
migration toward digital TV 168
MMDS (Wireless Cable) 80
mobile information device (MID) 36
Moore's law 5
Motion Picture Experts Group (MPEG) 56
multi-channel network 16
multicasting 124
multichannel microwave distribution system (MMDS) 57
multichannel multipoint distribution system (MMDS) 65
multimedia industry 8
multimedia market 1
multimedia metamarket 12
multimedia service 3, 98
multiplexing 58
multipoint microwave distribution system (MMDS) 59

N

near-video-on-demand (NVOD) 58
needs dimension 32
niche channels 139

O

offline digital TV 37
offline media 34
offline multimedia PC 37
one-way interactivity 100
online digital TV 37
online media 34
online multimedia PC 36
operating systems (OSs) 127

P

parental control 118
pay-per-view (PPV) 58, 61, 118
persistent content protection 190
personal video recording (PVR) 128
pixel 54
portability 76
protection of intellectual property 184

R

reception costs 91
regionality 76
registration 106
registrational interactivity 109
response time 97
return channel (RC) 70
return channel band 97
return path 70, 100
rights information storage 189
rights management processes 192

S

satellite 77
satellite operators 17
satellite platforms 16
set top box 67, 111
Shannon's law 33
smart cards 126
SMATV systems 61

standards 71
standards for video compression 56
subscriber management system (SMS) 126
substitute paradigm 25

T

t-commerce 128
technical vulnerability 76
technology 5
technology dimension 31
television brand 148
terrestrial broadcasters 16
terrestrial TV networks 62
theme channels 58
themed channel 139, 142
transmission 105
transmission costs 91
transmission of a digitised signal 54
transmission systems 59
transmissional interactivity 108
transponders 60
transport media technologies 34
TV banking 122
TV channels 15
TV industry 15
TV shopping 121
two-way interactivity 101

U

universal mobile telecommunication system (UMTS) 35
upstream 97

V

value chain 12, 189
video-on-demand (VOD) 128
viewers' cognitive system 144

W

wired network 34
wireless application protocol (WAP) 34
wireless cable 65
wireless network 34

Z

zapping 147

InfoSci-Online Database

30-Day free trial!

www.infosci-online.com

Provide instant access to the latest offerings of Idea Group Inc. publications in the fields of INFORMATION SCIENCE, TECHNOLOGY and MANAGEMENT

During the past decade, with the advent of telecommunications and the availability of distance learning opportunities, more college and university libraries can now provide access to comprehensive collections of research literature through access to online databases.

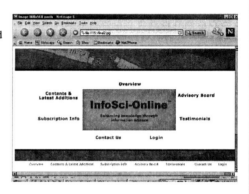

The InfoSci-Online database is the most comprehensive collection of *full-text* literature regarding research, trends, technologies, and challenges in the fields of information science, technology and management. This online database consists of over 3000 book chapters, 200+ journal articles, 200+ case studies and over 1,000+ conference proceedings papers from IGIís three imprints (Idea Group Publishing, Information Science Publishing and IRM Press) that can be accessed by users of this database through identifying areas of research interest and keywords.

Contents & Latest Additions:
Unlike the delay that readers face when waiting for the release of print publications, users will find this online database updated as soon as the material becomes available for distribution, providing instant access to the latest literature and research findings published by Idea Group Inc. in the field of information science and technology, in which emerging technologies and innovations are constantly taking place, and where time is of the essence.

The content within this database will be updated by IGI with 1300 new book chapters, 250+ journal articles and case studies and 250+ conference proceedings papers per year, all related to aspects of information, science, technology and management, published by Idea Group Inc. The updates will occur as soon as the material becomes available, even before the publications are sent to print.

InfoSci-Online pricing flexibility allows this database to be an excellent addition to your library, regardless of the size of your institution.

Contact: Ms. Carrie Skovrinskie, InfoSci-Online Project Coordinator, 717-533-8845 (Ext. 14), cskovrinskie@idea-group.com for a 30-day trial subscription to InfoSci-Online.

A product of:

INFORMATION SCIENCE PUBLISHING*
Enhancing Knowledge Through Information Science
http://www.info-sci-pub.com

*an imprint of Idea Group Inc.

New Release!

Multimedia Networking: Technology, Management and Applications

Syed Mahbubur Rahman
Minnesota State University, Mankato, USA

Today we are witnessing an explosive growth in the use of multiple media forms in varied application areas including entertainment, communication, collaborative work, electronic commerce and university courses. ***Multimedia Networking: Technology, Management and Applications*** presents an overview of this expanding technology beginning with application techniques that lead to management and design issues. The goal of this book is to highlight major multimedia networking issues, understanding and solution approaches, and networked multimedia applications design.

ISBN 1-930708-14-9 (h/c); eISBN 1-59140-005-8• US$89.95 • 498 pages • © 2002

"The increasing computing power, integrated with multimedia and telecommunication technologies, is bringing into reality our dream of real time, virtually face-to-face interaction with collaborators sitting far away from us."
– Syed Mahbubur Rahman, Minnesota State University, Mankato, USA

It's Easy to Order! Order online at www.idea-group.com or call our toll-free hotline at 1-800-345-4332!
Mon-Fri 8:30 am-5:00 pm (est) or fax 24 hours a day 717/533-8661

Idea Group Publishing

Hershey • London • Melbourne • Singapore • Beijing

An excellent addition to your library

New Release!

Enterprise Networking: Multilayer Switching and Applications

Vasilis Theoharakis
ALBA, Greece

Dimitrios Serpanos
University of Patras, Greece

The enterprise network is a vital component of the infrastructure of any modern firm. The adoption of networks in day-to-day and critical operations has made them indispensable, considering the large volume of information being shared and their penetration in almost every corporate activity and transaction. *Enterprise Networking: Multilayer Switching and Applications* addresses the technologies that have attracted significant attention and provides a promise for the scalability and future use in enterprise ìinformationî networks. More importantly, the book does not only cover these technologies, but also identifies and discusses other open issues that are currently being addressed.

ISBN 1-930708-17-3 (h/c); eISBN 1-59140-004-X • US$84.95 • 282 pages • Copyright © 2002

"Overall, *Enterprise Networking: Multilayer Switching and Applications* de-mystifies the hype and sorts through the masses of information presented by trade and technical journals by systematically presenting these issues across protocol layers."
–Vasilis Theoharakis, ALBA, Greece

**It's Easy to Order! Order online at www.idea-group.com or call our toll-free hotline at 1-800-345-4332!
Mon-Fri 8:30 am-5:00 pm (est) or fax 24 hours a day 717/533-8661**

Idea Group Publishing

Hershey • London • Melbourne • Singapore • Beijing

An excellent addition to your library